CONTENTS

PREFACE

Interest in the phenomenon of superplasticity and its commercial potential as the basis of an important forming process has significantly increased in recent years and this book is written as an introduction to this rapidly expanding subject. It is aimed at final year undergraduate and graduate students in metallurgy, materials science and materials engineering, but should also prove useful to those in industry concerned with superplastic forming/diffusion bonding technology. The book has been developed from a series of lectures offered at the Materials Science Centre, University of Manchester.

The topics covered include: the main characteristics of superplastic materials; processing to develop superplastic microstructures in a range of materials; experimental techniques to assess the extent of superplasticity; mechanisms of superplastic deformation; cavitation during superplastic flow; superplastic forming and diffusion bonding. Where appropriate the concepts presented are accompanied by worked examples. It has not been possible to cover the topic of superplastic forming as completely as would have been desirable since much of this field is now based in engineering mechanics and, in particular, finite element modelling, the details of which are beyond the scope of this text.

John Pilling
Norman Ridley

July 1988

Cover photos: Aluminium bronze (Cu-10Al-5Fe-5Ni) showing as-cast (optical) and thermomechanically processed (SEM) microstructures consisting of copper-rich solid solution and aluminide phases; the processed material is exceptionally super-plastic, with recently reported tensile elongations of greater than 8000% (K. Higashi, private communication), and can be readily bulge-formed.

1
AN INTRODUCTION TO SUPERPLASTICITY

1.1 Background

Superplastic materials are polycrystalline solids which have the ability to undergo large uniform strains prior to failure. For deformation in uniaxial tension, elongations to failure in excess of 200% are usually indicative of superplasticity, although several materials can attain extensions greater than 1000%. The highest elongations reported are 4850% in a Pb-Sn eutectic alloy, Fig. 1.1 [1] and greater than 5500% for an aluminium bronze [2].

Observations of what appeared to be superplastic behaviour were made in the late 1920s by Hargreaves [3,4] and Jenkins [5]. However, the most spectacular of the earlier observations was that by Pearson in 1934 [6]. While working on eutectics, he reported a tensile elongation of 1950% without failure for a Bi-Sn alloy. Pearson also examined the bulging characteristics of his materials using internally pressurised tubular specimens. Following these observations there was little further interest within the Western world in what was clearly regarded as a laboratory curiosity. Nevertheless, studies were carried out in the USSR and the term superplasticity was coined by Bochvar and Sviderskaya in 1945 to describe the extended ductility observed in Zn-Al alloys [7]. (The term superplasticity is derived from the Latin prefix 'super' meaning excess and the Greek word

1

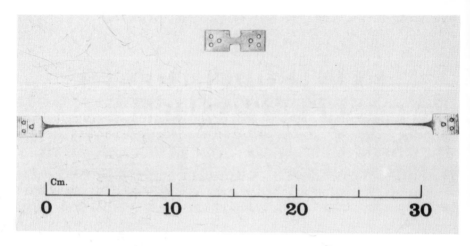

Figure 1.1 Exceptional superplasticity in Pb-62%Sn eutectic pulled in tension at 413K to an elongation of 4850% [1].

'plastikos' which means to give form to.) In 1962 Underwood [8] reviewed previous work on superplasticity and it was this paper, along with the subsequent work of Backofen and his colleagues [9] reported in 1964, which was the forerunner of the present expanding scientific and technological interest in superplasticity.

There are two main types of superplastic behaviour: micrograin or microstructural superplasticity and transformation or environmental superplasticity. Both types of behaviour have been reviewed by Padmanabhan and Davies [10].

These authors examined the experimental methods of assessing the superplastic deformation characteristics of materials. They also summarised the observed features of superplasticity and compared them with the many hypotheses presented in the literature up to 1980. The present volume will concentrate almost exclusively on micrograin superplasticity since this has been the most extensively studied and is of considerable significance from a commercial viewpoint.

Micrograin superplasticity is shown by materials with a fine grain size, usually less than 10 μm, when they are deformed within the strain rate range 10^{-5} to 10^{-1}/s at temperatures greater than $0.5T_m$, where T_m is the melting point in degrees Kelvin. Superplastic deformation is characterised by low flow stresses and this, combined with the high uniformity of plastic flow, has led to considerable commercial interest in the superplastic forming of components using techniques similar to those developed for bulge forming of thermoplastics. Superplastically formed parts find many uses particularly in aerospace. Superplastic forging of nickel-base alloys has been used to form turbine discs with integral blades, while diffusion bonding and superplastic forming (DB-SPF) of titanium alloys is used to produce fan and compressor blades for aeroengines. Aluminium alloys can be used in the fabrication of airframe control surfaces and small-scale structural elements where low weight and high stiffness are required. Non-aerospace applications of Al alloys include containers with complex surface profiles and decorative panels for internal and external cladding of buildings.

1.2 Mechanical Aspects of Superplasticity

The most important mechanical characteristic of a superplastic material is its high strain rate sensitivity of flow stress, referred to as 'm' and defined by

$$\sigma = k\dot{\epsilon}^m \qquad (1.1)$$

where σ is the flow stress, $\dot{\epsilon}$ the strain rate and k a material constant. If superplasticity is regarded as a type of creep behaviour then equation (1.1) may be rewritten as

$$\dot{\epsilon} = k'\sigma^n \qquad (1.2)$$

where n, the stress exponent for deformation is

equal to 1/m. For superplastic behaviour, m would be greater than or equal to 0.3 and for the majority of superplastic materials m lies in the range 0.4 to 0.8. The presence of a neck in a material subject to tensile straining leads to a locally high strain rate and, for a high value of m, to a sharp increase in the flow stress within the necked region. Hence the neck undergoes strain rate hardening which inhibits its further development. Thus a high strain rate sensitivity confers a high resistance to neck development and results in the high tensile elongations characteristic of superplastic materials.

It can be seen from equation (1.1) that if the relationship between σ and $\dot{\epsilon}$ is measured and then plotted logarithmically the slope of the plot is equal to the strain rate sensitivity of the flow stress, m. In practice most superplastic materials show a sigmoidal variation of the flow stress with strain rate. In Fig. 1.2 it is evident that the strain rate sensitivity passes through a maximum. A value of m>0.3 delineates the superplastic regime (Region II). Both the high and low strain rate ranges exhibit values of m in the range 0.1 to 0.3. The region at high strain rates (Region III) is generally believed to correspond to conventional recovery controlled dislocation creep (power law creep). Deformation within this region leads to the observation of slip lines and to the development of high dislocation densities within the grains. Crystallographic texture within the material is increased and significant grain elongation occurs during deformation.

In the superplastic regime, Region II, where high uniform strains are observed, experimental studies have so far failed to identify a unique rate controlling mechanism of deformation. It is clear, however, that grain boundary sliding and grain rotation make a substantial contribution to the total strain. In contrast to region III the grains remain equiaxed throughout deformation and materials which initially show microstructural banding develop a more uniform equiaxed microstructure. Crystallographic texture may be

4

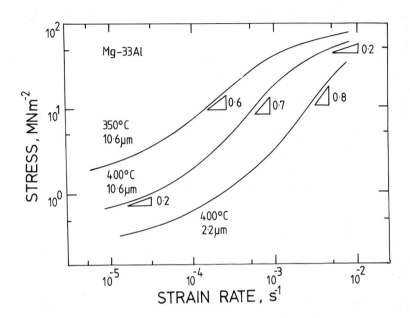

Figure 1.2 Measured variation of the flow stress with strain rate for a superplastic Al-Mg eutectic at different temperatures and grain sizes [16].

reduced during deformation in this region. Transmission electron microscopy studies have shown only limited evidence for dislocation activity within the grains of superplastically deformed materials. The flow stress, σ, decreases and the strain rate sensitivity, m, increases with increasing temperature,T, and decreasing grain size, d, (Fig. 1.2). The elongation-to-failure in this region tends to increase with increasing m [11].

The origin of the low strain rate regime (Region I) is at present unknown. The experimental evidence available at these low strain rates is limited and often contradictory [12]. It has been suggested that the decrease in the strain rate sensitivity with decreasing strain rate is only apparent and results from a threshold stress for deformation, or from the effects of micro-structural instability (grain growth hardening). Alternatively, the similarity in stress exponent ($n=1/m$) between Regions I and III has been used to imply that Region I also involves recovery controlled dislocation creep. Other experimental investigations, however, have shown that at low strain rates the strain rate sensitivity can increase, taking values close to unity and thereby implying diffusion creep. The experimental evidence relating to Region I will be considered in more detail in Chapter 4.

The mechanical behaviour of superplastic materials is very sensitive to both temperature and grain size. In general, increasing the temperature or decreasing the grain size of the material has a similar effect on the variation of flow stress with strain rate (Fig. 1.2). Increasing the temperature decreases the flow stress, particularly at the lower strain rates corresponding to the transition from Region II to Region I. The maximum strain rate sensitivity has been found to increase with increasing temperature and the strain rate of maximum 'm' moves to higher strain rates. The increases in 'm' values are much greater in Region II than in Region III. The strain rates at which superplasticity is normally observed lie in the range 10^{-5} to 10^{-1}/s, although this is more usually between 2×10^{-4} and 2×10^{-3}/s. These strain rates are less than those used in conventional hot deformation processes.

1.3 Types of Superplastic Materials

For superplastic behaviour a material must be capable of being processed into a fine equiaxed grain structure which will remain stable during deformation. The grain size of superplastic materials is normally in the range 2 to 10 μm, although a limited amount of superplasticity is still observed for grain sizes up to 20 μm or even higher. In the presence of a suitable microstructure, superplasticity occurs over a

6

narrow range of temperatures which generally lies above $0.5T_m$.

There are two main types of superplastic alloys: pseudo single phase and microduplex. In the former class of material, a combination of hot and cold working and heat treatment is employed to develop a fine scale distribution of dispersoids so that on recrystallisation the alloy will have a grain size of the order of 5 μm or less. Ideally, the dispersion of particles will also prevent any further grain growth during superplastic deformation. The precipitation strengthened aluminium alloys based on the 2xxx series (Al-Cu), 7xxx series (Al-Zn-Mg) and the (8xxx) series Al-Li alloys can be included in this group. Other materials include dispersion strengthened copper alloys where silicide or aluminide particles are used to limit grain growth. Some steels and ultra-fine grain ceramics such as UO_2, can also be classed as essentially single phase. The aluminium alloys, which from a commercial viewpoint are the most important of the pseudo single phase materials, can be further subdivided into those which are recrystallised prior to superplastic forming and those which acquire their fine grain structure only after a limited amount of deformation at the forming temperature.

The microduplex materials are thermomechanically processed to give a fine grain or phase size. Grain growth is limited by having a microstructure that consists of roughly equal proportions of two or more chemically and structurally different phases. This latter group of materials includes α/β titanium alloys, α/τ stainless steels, α/β copper alloys, eutectics and some ceramics. In the case of the ceramic materials, tensile elongations greater than 200% have been reported. Although these elongations are small by comparison with those commonly attained in metallic materials, they are much larger than the 1 to 2% normally observed in structural ceramics [13-15].

2
SUPERPLASTIC MATERIALS

2.1 Introduction

It is well established that a fine grain size is an essential prerequisite for superplasticity. An understanding of the basic metallurgical principles underlying grain refinement and grain growth is therefore important to the development of superplasticity in materials which would not normally be superplastic. Several methods are available for grain refinement including phase separation, phase transformation, and mechanical working with recrystallisation. It should be possible in principle to develop fine grain microstructures using thermal treatments alone. However, the imposition of mechanical working during any stage of the heat treatment may give a thermomechanical process which produces the required grain size in a fewer number of processing steps.

Unfortunately, merely achieving a fine grain size is not in itself sufficient to guarantee that a material will exhibit superplasticity, since the grain size needs to remain stable throughout the deformation process. Grain growth during superplastic flow has been reported for a number of materials [17-28] with the extent of grain growth being greater in the superplastically deformed part of the samples studied than in the undeformed areas. From the available data it is clear that strain enhanced grain growth is a widespread property of superplastic deformation in both pseudo single phase and microduplex materials.

In this Chapter the metallurgical principles underlying the thermomechanical and thermal treatments that are capable of refining the grain size of a material are described. This is then followed by a more detailed account of the procedures that have been applied to develop superplasticity in aluminium, iron and titanium based alloys. The Chapter concludes with a summary of the observations of superplastic behaviour in these and other materials, including ceramics.

2.2 Grain Refinement by Mechanical Working

2.2.1 Duplex alloys

Grain refinement in equilibrium duplex alloys is accomplished by hot working the material close to the temperature range where superplastic deformation is to be carried out. In most cases the temperature is as high as possible concomitant with a microstructure consisting of approximately equal volume fractions of the two phases. If the phases have different deformation characteristics, such as one being harder and more brittle than the other, then working would fragment the harder phase. The softer phase would then be forced to infiltrate and separate the harder phase. The phases can recrystallise or spheroidise during the hot working process (Fig. 2.1). Alternatively, if the two phases have very similar mechanical responses, working first elongates the grain structure and then fragments it by the development of intense shear bands. Again the original structure can be reformed, but on a much finer scale, by recrystallisation. In some cases, the fine equiaxed structure is developed from a cold rolled material by recrystallisation during heating immediately prior to superplastic deformation.

Grain growth during both heat treatment and superplastic flow is restricted as the individual phases have different chemical compositions. In

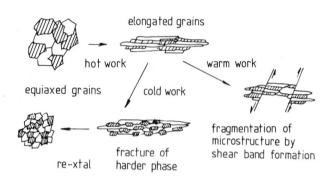

Figure 2.1 Schematic illustration of the microstructural changes that occur during grain refinement by mechanical working of duplex alloys.

order that grain growth can occur solutes must diffuse from the smaller grains of each phase through or around the other phase to the larger grains. As alloy additions partition to the phase in which they are most soluble, the rate of mass transfer of the alloy element through the other phase is restricted by its limited solubility. The grain size is then said to be 'segregation stabilised'. A duplex superplastic material of very high structural stability would be one in which the component metals (or ceramics) showed little or no solid solubility.

Examples of duplex materials which can be hot worked to develop fine grain microstructures include α/β titanium alloys [29,30] α/β copper alloys, ultra-high carbon steels [31-33], α/τ duplex stainless steels and eutectics such as Al-Ca [34], Al-Ca-Zn [35], Pb-Sn and Bi-Sn [6].

2.2.2 Pseudo single phase alloys

Grain refinement as a result of warm working and recrystallisation treatments has been used extensively in the development of superplastic

10

aluminium based alloys. The alloys are termed pseudo single phase since they consist almost entirely of a solid solution strengthened matrix with <10% by volume of a precipitate phase which is present to stabilise the microstructure against grain growth. The alloys are constituted such that two types of particles are present during warm working but that usually only one, the finer of the two, remains at the superplastic forming temperature.

The high density of dislocations which form during warm working would normally tend to rearrange themselves by climb to form dislocation walls and possibly subgrain boundaries. In the absence of any fine particles the dislocation arrays would migrate by climb allowing the microstructure to undergo continuous recovery and ultimately recrystallisation. However, the presence of fine particles, which are usually less than 0.2 μm in diameter, prevents recovery by exerting a drag on the migrating dislocations, dislocation walls, and subgrain boundaries. Mechanical working in the presence of the fine particles therefore generates and maintains a large amount of stored energy and introduces into the microstructure a large number of potential nucleation sites for subsequent recrystallisation.

Alloys which contain predominantly fine particles develop a fine grain equiaxed microstructure during the initial stages of superplastic deformation by in situ recrystallisation. The particles and solutes which prevent recrystallisation during warm working at lower temperatures cease to be effective in restraining dislocation migration at the higher deformation temperatures. The majority of the grains which form on heating to the superplastic deformation temperature have very similar crystallographic orientations because of the texture introduced into the material during warm working. As high temperature deformation and hence grain boundary

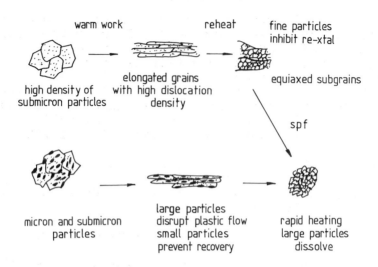

Figure 2.2   Schematic illustration of the microstructural changes that occur during grain refinement by mechanical working of pseudo single phase alloys.

sliding (+) proceeds, the misorientation between the grains increases and leads to the formation of high angle grain boundaries and a true superplastic microstructure which can undergo grain boundary sliding (Fig. 2.2).

A second group of pseudo single phase alloys contains both fine and coarse particles during processing. The presence of larger particles, which are usually greater than 1 μm in diameter, provide nucleation sites for recrystallisation and cause localised distortions in the orientation of plastic flow during warm working. The resulting

(+) For the moment it will be assumed that grain boundary sliding is the mechanism by which superplastic flow occurs. The validity of this assumption will be examined in detail in Chapter 4.

differences in crystallographic orientation of the material, on a very fine scale, lead to the nucleation of recrystallised grains with widely varying orientations and the formation of high angle boundaries as the embryonic grains impinge. Materials which contain predominantly the larger particles are statically recrystallised to produce the fine equiaxed microstructure by annealing prior to superplastic deformation (Fig. 2.2).

Alloys which are refined by recrystallisation include Al-Cu-Zr (2xxx series alloys including Supral), Al-Mg-Zr (5xxx series), Al-Zn-Mg-Cr (7xxx series) and Cu-Al-Si-Co (Coronze).

2.3 Grain Refinement by Phase Transformation

Several studies have shown that repeated thermal cycling of a material through a phase transformation can result in a very fine grain size. The mechanism of grain refinement is the nucleation of the reaction product at several sites on the grain boundaries of the parent phase. The product phase then grows as the transformation proceeds, replacing the single parent grains by a multitude of smaller grains. Repeated cycling through the phase transformation further refines the structure until a saturation grain size is reached (Fig. 2.3). A similar effect can be achieved by controlled rolling above the transformation temperature such that a heavily deformed but unrecrystallised parent phase is produced. The high dislocation density within the parent phase results in a large number of nucleation sites for the product phase. The parent phase transforms directly to a fine grain product on cooling. Ideally superplastic deformation would take place in the temperature range where the two phases are present in order that the grain size is stabilised. However, if superplastic deformation is to be carried out at a temperature where the material is single phase then additional steps need to be taken to prevent grain growth - e.g. the introduction of fine dispersoids to pin the grain boundaries.

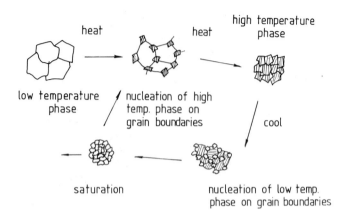

heat   heat   high temperature phase

low temperature phase   nucleation of high temp. phase on grain boundaries   cool

saturation   nucleation of low temp. phase on grain boundaries

Figure 2.3 Schematic illustration of the microstructural changes that occur during grain refinement by phase transformation on thermal cycling.

Controlled rolling coupled to phase transformation has been successfully exploited in high strength low alloy steels (HSLA), ultra-high carbon steels [26] and could equally be applied in the Cu-Al based alloys where the eutectoid transformation ($\beta \rightarrow \alpha + \tau_2$ or $\beta \rightarrow \alpha + K_{III}$) occurs on cooling [2].

2.4 Grain Refinement by Phase Separation

It is often possible to anneal duplex materials at a temperature where only one phase is stable. Quenching the resulting single phase structure will then either effect a martensitic transformation or produce a supersaturated solid solution. A subsequent annealing treatment will result in the separation of the two equilibrium phases from the metastable microstructure. If sufficient nucleation sites are available then a fine grain microstructure is produced. A martensitic structure, such as that formed on quenching α/β titanium alloys from the β phase field, provides such a density of potential nucleation sites for the α grains and hence aids the change of the β'-Ti

14

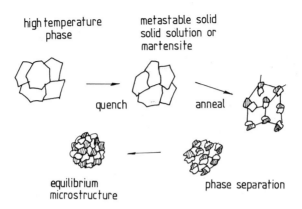

Figure 2.4   Schematic illustration of the microstructural changes that occur during grain refinement by phase separation from a non-equilibrium microstructure.

to the equilibrium α+β phases, (Fig 2.4). Another example of grain refinement by phase separation is the spinodal decomposition of Zn-22%Al. Slow cooling the Zn-Al alloy results in a lamellar eutectoidal decomposition product with very poor superplastic properties [36]. If however, the alloy is quenched to room temperature a supersaturated solid solution of Al in Zn is formed which decomposes in a spinodal fashion producing a very fine grain, highly superplastic, microstructure [37,38].

Despite the wide variety of methods available for developing fine grain microstructures, only a very small number of upwards of 100 distinct alloys which show extensive superplasticity are, or have the potential to be, exploited on a commercial scale [10]. These include medium to high strength aluminium alloys, some duplex titanium alloys and ultra-high carbon and stainless steels.

Table 2.1 Summary of the conditions under which superplastic behaviour is shown by aluminium-base alloys. (Upper section: duplex alloys: Lower section: pseudo single phase alloys)

Alloy	Temperature °C	Strain rate %/min	m	Elongation %	Ref.
Al-33Cu	480-520	24 - 120	0.6 - 0.8	400 - 1000	43,44
	470	6 - 120	0.5	600 - 1200	46
Al-5Ca-5Zn	450-525	120	0.4	600	35
Al-6Cu-0.4Zr-0.2Mg (Supral 100)	420-480	3 - 24	0.4 - 0.55	800 - 1200	45,47
Al-6Cu-0.4Zr-0.3Mg-0.2Si-0.1Ge (Supral 220)	460	4 - 24	0.65	>1800	48
Al-5Mg-0.6Cu-0.7Mn-0.15Cr (Neopral)	480-530	6	0.45 - 0.7	700	49
Al-6Mg-0.4Zr	520	1.2	0.6	885	50
Al-6Zn-3Mg	320-360	6 - 60	0.3 - 0.35	200 - 400	51
Al-9Zn-1Mg-0.3Zr	550	-	0.9	1550	52
Al-5.5Zn-2.5Mg-1.5Cu-0.2Cr (7475)	515	1.2 - 5	0.5 - 0.8	1200	53,54
Al-6.2Zn-2.5Mg-1.7Cu (7010)	520	0.18 - 0.5	0.65	>350	55
Al-2.5Li-1.2Cu-0.6Mg-0.1Zr (8090)	500-540	0.6 - 6	0.4 - 0.55	500 - 1000	56
Al-3Li-0.5Zr	425-475	12 - 60	0.45	500 - 1000	57,58
Al-4Cu-3Li-0.5Zr	450	12 - 180	0.45	500 - 800	57,58

2.5 Aluminium alloys

Of the aluminium alloys which have either been specially developed or processed for super-plasticity, only two are used extensively in structural applications, Al-7475 and Supral [39] (including Supral 100, Supral 150 and Supral 220), although there is substantial interest in the use of aluminium-lithium alloys such as Al-8090. Alloys such as Supral 5000 (Al-2Mg-0.4Zr), Formal 548 and Neopral (Al-5Mg-0.15Cr) are used for decorative panels in architectural applications. The compositions of a number of pseudo single phase and microduplex aluminium alloys which have been shown to be superplastic are given in Table 2.1.

2.5.1 Dynamically recrystallised alloys

In the Al-Cu-Zr alloys such as Supral 100 and Supral 220, a fine grain structure is achieved by in situ recrystallisation during the early stages of superplastic deformation rather than immediately following warm working [40,41]. The processing route for Supral 220 is shown schematically in figure 2.5. During the initial homogenisation of the cast alloy, metastable cubic $ZrAl_3$ particles are precipitated and most of the copper and magnesium is taken into solution. The $ZrAl_3$ particles have a volume fraction of ~5% and are distributed homogeneously with a particle size of between 5 and 10 nm.

During warm working at 300°C, recovery and recrystallisation of the structure is prevented by the $ZrAl_3$ particles, and to a lesser extent by the Mg_2Si particles and the copper and magnesium in solid solution, Fig. 2.5. During the early stages of superplastic deformation at 460°C the structure dynamically recrystallises. Recrystallisation occurs 'in situ', the low angle subgrain boundaries which form on heating to the super-plastic deformation temperature become progressively more high angle and thus able to contribute to grain boundary sliding (Figs. 2.5 and 2.6).

Tensile elongations in excess of 1500% are commonly attained in Supral 220 with a strain rate.

sensitivity of 0.5-0.6 at 460°C. Moreover, the high elongations are obtained at strain rates between 10^{-3} and 10^{-2}/s and thus superplastic forming can be carried out quickly, minimising both forming times and the deleterious effects of strain hardening due to grain growth. Indeed there can be advantages to forming at high strain rates during the early stages of deformation to ensure that a very fine grain size is produced, since the recrystallised grain size decreases with increasing initial strain rate [42]. The forming can then be completed at a lower, more optimum, strain rate for superplasticity. However, the presence of coarse particles of $CuAl_2$ and primary $ZrAl_3$, which tend to be clustered in bands within the processed sheet, can act as sites for cavity nucleation during superplastic flow (see Chapter 5). The extent to

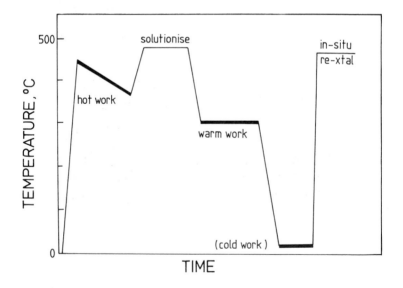

Figure 2.5 Schematic processing route for Supral 220 used to generate a fine grain microstructure by 'dynamic' or 'in-situ' recrystallisation.

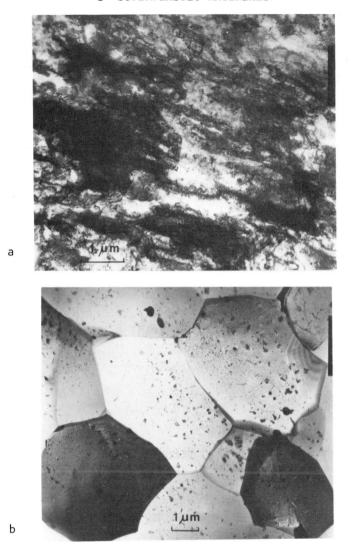

a

b

Figure 2.6 Microstructure of Supral in (a) the
as-rolled condition and (b) after dynamic
recrystallisation [40].

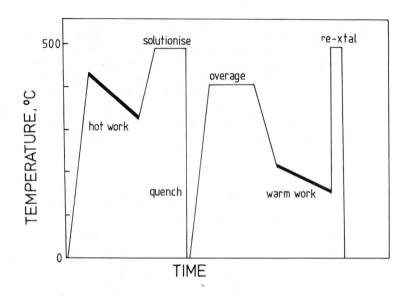

Figure 2.7 Schematic processing route for Al-7475 used to generate a fine grain microstructure by 'static' recrystallisation.

which cavitation damage may develop during superplastic flow can, if it is not controlled by other means, be quite high and may often approach 10% by volume at high strains. However, at commercial levels of strain, typically $\epsilon < 1.2$, cavitation is more likely to be around 1% by volume.

2.5.2 Statically recrystallised alloys

High elongations to failure are also exhibited by Al-Zn-Mg-Cr alloys such as 7475 [53]. However, unlike Al-Cu-Zr alloys, the 7xxx series alloys require a more complex thermomechanical processing treatment to develop a fine grain size in what is not normally a superplastic material. (A typical processing schedule for 7475 is illustrated in figure 2.7.) A solution heat treatment consisting of 3 h at 480°C is used to dissolve all but the Cr-rich dispersoids. The latter particles, probably

20

$Mg_3Cr_2Al_{18}$, $CrAl_7$, or Cr_2Al_9, are insoluble and typically 0.1-0.2 μm in diameter. The alloy is then annealed at 400°C for 8 h to produce an overaged dispersion of M phase (a mixture of $MgZn_2$ and CuMgAl) and T phase (a mixture of $Mg_3Zn_3Al_2$ and $CuMg_4Al_6$) precipitates with diameters between 1 and 2 μm. The latter particles have two functions. Firstly, to effect inhomogeneous deformation during warm rolling and secondly to act as nucleation sites for the recrystallising grains during subsequent heat treatment.

The material is warm rolled at around 200°C, following overageing, creating areas of intense deformation and substantial lattice reorientation around the undeformable M phase precipitates (Fig. 2.8a). After warm working, a high temperature annealing treatment is applied to produce a fully recrystallised microstructure. The fine scale dispersion of large precipitates and a rapid heating rate result in a large number of recrystallisation nuclei, and thus a small grain size is developed. Grain growth during annealing and subsequent superplastic deformation is restricted by the drag effect imposed on the grain boundaries by the fine insoluble Cr-rich dispersoid phase [54,59] (Fig. 2.8b).

Similar processes have been developed for other Al-Zn-Mg alloys including 7010, 7050 and 7075 and Al-Li alloys with Zr, Mg and Cu additions. In the Al-Li alloys grain boundary pinning is provided by a fine dispersion of submicron sized particles of $ZrAl_3$. The alloys are solution treated at around 550°C for 3 h, water quenched, then overaged at 510°C to develop large particles of T phase (Al_2LiCu (T_1) and Al_5Li_3Cu (T_2)) and S phase ($CuMgAl_2$). After warm working, a rapid recrystallisation heat treatment is carried out at 550°C to produce the fine grain size necessary for superplasticity.

2.5.3 Powder metallurgy alloys

Al-Li-Zr based alloys containing between 2 and 4 wt-% lithium have been converted into powders by

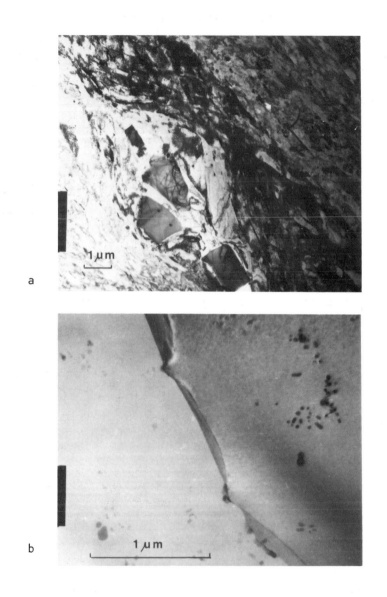

a

b

Figure 2.8 Microstructure of Al-7475 in (a) the
warm rolled condition showing the non-uniform plastic
flow around the coarse intermetallic particles and
(b) the pinning of the recrystallised grain
boundaries by the fine intermetallic particles
[40].

rapid solidification/atomisation techniques. The powders can be used as a precursor material for the production of slab and ultimately sheet alloys. The powders are generally cold compacted, degassed and then hot pressed in vacuum. The compact is then extruded at around 400°C to give a fully dense material. Although the extruded slabs have a fine grain size, they are not normally superplastic as the grain boundaries are predominantly of a low angle type. Additional thermomechanical processing, such as that described in Section 2.5.2, is usually required in order to develop a fully superplastic microstructure [58,60]. The powder metallurgy route for the production of Al-Li based alloys may be preferred to the normal ingot metallurgy route as a more uniform product with a higher lithium content can be produced.

The direct powder route may however prove impractical for producing large volumes of material since fine aluminium powders are highly explosive and need special handling facilities. An alternative method of producing larger preforms is that involving metal spraying where a jet of molten alloy is atomised in an inert environment, and then condensed onto a semisolid surface. The preform is built up rapidly as additional semimolten droplets of metal are forced into contact with the surface of the preform itself.

The spray cast billet can then be upset forged or extruded to both consolidate and remove any porosity which developed as the billet was being built up. This treatment may then be followed by warm rolling to around 90% reduction to give an alloy of the dynamically recrystallising type (e.g. Al-4Cu-3Li-0.5Zr) or thermomechanical processing to give a fine grain material by the static recrystallisation route described in the previous section.

2.5.4 Eutectic alloys

Several aluminium alloy systems solidify in a rodlike morphology including $Al-CuAl_2$ [19,44] $Al-Si$ [61], $Al-CaAl_4$ [34] and $Al-Mg_2Si$ [62], but

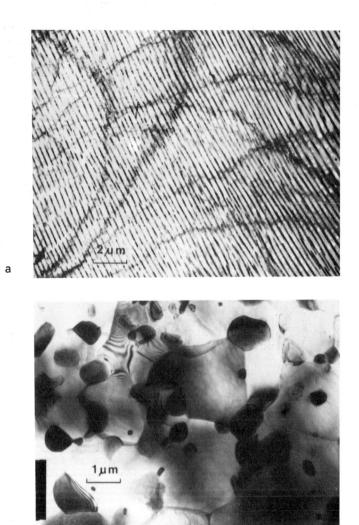

a

b

Figure 2.9 Microstructure of Al-5Ca-5Zn in (a) the
as-cast condition and (b) after hot working [40].

only the ternary Al-Ca-Zn eutectic has been considered
to have commercial potential [35]. The Al-5Ca-5Zn alloy
can be chill cast to produce a lamellar structure with
a cell size of ~10 μm, in which the Al-CaZnAl$_3$ inter-
lamellar spacing is between 0.2 and 0.4 μm, (Fig.
2.9a). The lamellar structure can be broken up by
hot and cold rolling to give partially spheroidised
particles of CaZnAl$_3$. On heating to about 450°C a
fine grain structure forms by in situ recrystallisation of the aluminium matrix (Fig. 2.9b). A grain
size in the range 1 to 2 μm has enabled strain rate
sensitivities greater than 0.5 to be recorded at
deformation temperatures between 500 and 550°C.
Elongations to failure of the order of 600% have
been obtained at strain rates of 2 x 10^{-2}/s. Grain
growth during superplastic flow is minimal because
of the high volume fraction of intermetallic
particles present in the microstructure.

2.6 Iron alloys

Iron based alloys are probably the most versatile
and, in many contexts, the most important of all
structural materials. It is not surprising,
therefore, that superplasticity has been developed
in a number of these alloys. The compositions and
properties of several superplastic iron based
alloys are given in Table 2.2, and also in the
reviews by Ridley [63] and by Walser [64]. The
various alloys can be divided into two main types:
particle stabilised and segregation stabilised.

2.6.1 Particle stabilised alloys

This group consists of primarily plain carbon and
low alloy steels. The fine grain microstructure
required for superplasticity can be developed by
thermal and thermomechanical processing and is
stabilised mainly by carbide particles.

For steels containing up to eutectoid carbon
content, finely dispersed cementite particles in a
matrix of small ferrite grains can be developed by
austenitising, quenching to produce martensite,
followed by tempering at 550 to 700°C. However,
during deformation at 700°C, the cementite

Table 2.2 Summary of the conditions under which superplasticity is observed in iron-base alloys. (Upper section - particle stabilised alloys: Lower section - segregation stabilised alloys)

Alloy	Temperature °C	Strain rate %/min	m	Elongation %	Ref.
Fe-(0.2-0.8)C	705	0.1	0.35	98	70
Fe-0.8C	716	0.4	0.42	133	24
Fe-1C-1Mn-0.5Cr-0.5W-0.2V	650	0.06 - 6	0.3 - 0.5	150 - 1200	65
Fe-1.3C-1.5Cr-3Si	800	0.3 - 12	0.5	-	64
Fe-(1.3-1.9)C	650	0.6 - 60	0.35 - 0.5	200 - 1500	33
Fe-1.6C-1.5Cr	750 - 850	0.6 - 0.6	0.45 - 0.55	800	27,33
	650	0.6 - 6	0.4 - 0.45	1100 - 1500	28
Fe-4Ni-3Mo-1.6Ti	900 - 960	6	0.6 - 0.8	400 - 1000	71
Fe-26Cr-5Ni-3Mo (Ferralium 255)	950	6	0.45 - 0.5	600	72
Fe-1.9Mn-0.4C	727	1.8	0.5	460	73
Fe-26Cr-6.5Ni-0.4Ti (IN-744)	960	18	0.55	>1000	66
Fe-25Cr-6.5Ni-3Mo-0.14N	900 - 1050	2.4 - 6	0.35 - 0.5	900 - 2500	69
Fe-18.5Cr-5Ni-3Mo (Avesta 3RE60)	1000	6	0.7	>700	67
Fe-22Cr-5.5Ni-3Mo-2Mn (SAF 2205)	900 - 1000	0.6 - 6	0.3 - 0.4	500 - 1000	68

particles coarsen rapidly leading to grain growth and to poor superplastic behaviour. Cavities nucleate at the ferrite/cementite interfaces and contribute to premature failure [24,64,65].

In an attempt to circumvent this problem, steels with carbon contents in the range 1.2 to 2.1% have been examined. It was reasoned that the higher carbon contents would increase the volume fraction of cementite and would be more effective in restraining grain growth. Moreover, there would also be the possibility of extending the deformation temperature above the A_1 into the τ + Fe_3C phase field (Fig. 2.10). A typical treatment involves homogenisation of the austenite in the range 1100-1150°C to take all of the carbon into solution, followed by rolling during cooling through the τ + Fe_3C region to just below the A_1 temperature. The rolling reductions are about 10% per pass to give a total true strain of about -1.5. Isothermal rolling to a further true strain of -1.5 is then applied to spheroidise the pearlite which forms. The total true strain involved is approximately -3.0. Figure 2.11 shows the microstructure of an ultra-high carbon steel in the as-cast condition and after thermomechanical processing.

Tensile elongations to failure in ultra-high carbon steels (UHC) have ranged from as low as 50% to values in excess of 1500% at 650°C depending on the carbon content, processing history and purity of the steel. The lower elongations and lower 'm' values were observed in high purity alloys while the higher elongations together with 'm' values between 0.3 and 0.5 were found in commercial purity material. Subsequent work involving the addition of up to 1.5 wt-% Cr to the steels was found to enhance the elongation to failure and to improve microstructural stability, with virtually no strain (grain growth) hardening being observed. The chromium along with impurities and alloy additions such as Mn, Ti,V and W, retard the coarsening kinetics of the carbide particles and hence help to maintain a stable microstructure during superplastic deformation [27,28,32,65].

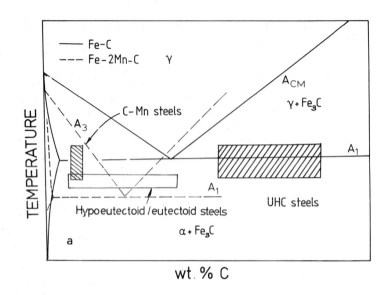

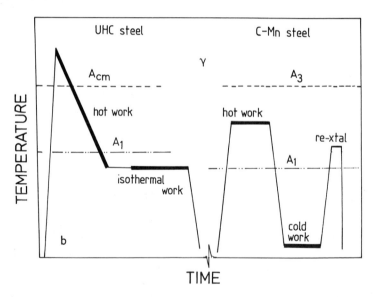

Figure 2.10 (a) Fe–C and Fe–2Mn–C phase diagrams
showing the regions of composition and temperature
where superplasticity would be expected. (b) Schematic
processing routes used to generate a particle
stabilised microstructure in C–Mn and UHC steels.

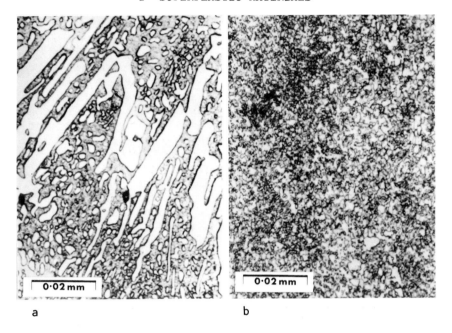

a   b

Figure 2.11 Microstructure of ultra-high-carbon
(UHC) steels in (a) the as-cast condition and (b)
after thermomechanical processing [26]

2.6.2 Segregation stabilised alloys

This group includes medium carbon low alloy
steels, C-Mn steels and microduplex stainless
steels, such as IN-744 (Fe-26Cr-6.5Ni-0.4Ti) and
Avesta 3RE60 (Fe-18.5Cr-3Mo-5Ni), which are
essentially carbon free. The fine grain
superplastic microstructure is developed either by
thermal cycling or formed during the conventional
hot and cold rolling schedules used to produce
sheet material from ingots. The microstructure
during superplastic deformation normally consists
of roughly equal volume fractions of α and τ
phases. Since the various alloy additions partition

preferentially to one or other of the two phases, the microstructure is segregation stabilised.

In principle, plain carbon steels of medium carbon content deformed in the $\alpha + \tau$ phase field should exhibit superplasticity. Fine grain microstructures consisting of approximately equal volume fractions of the two phases can be obtained by cold working followed by recrystallisation in the $\alpha + \tau$ region. However, grain growth at deformation temperatures between 750 and 850°C is rapid. Additions of up to 2% Mn improve the superplastic properties. The manganese segregates to the austenite phase and helps to stabilise the grain size as manganese diffusion is much less rapid than that of carbon. In addition, manganese lowers the A_1 enabling deformation to be carried out at lower temperatures in the two phase field, (Fig. 2.10a). Nevertheless, elongations to failure in C-Mn steels deformed in the $\alpha + \tau$ field are much lower than those obtained in either the $\alpha + Fe_3C$ or $\tau + Fe_3C$ phase fields (see Table 2.2).

Several stainless steels are commercially available in sheet form with a fine grain duplex microstructure. The steels contain a balance of ferrite stabilisers (Cr,Ti,Mo) and austenite stabilisers (Ni,Mn,N) which result in a two phase microstructure which is stable from room temperature to >1000°C (Fig. 2.12). The duplex microstructure is developed by hot deformation from 1200 to 1250°C in the ferrite phase field followed by annealing at 900°C to precipitate austenite, or by hot deformation, cold work and recrystallisation at 900°C to give a fine grain equiaxed microstructure (Fig. 2.12). Commercially available materials include IN744, Avesta 3RE60, SAF 2205, all of which are capable of elongations to failure in excess 1000% [66-68].

Maehara and Ohmari [69] examined superplasticity in a duplex stainless steel of composition Fe-25Cr-6.5Ni-3Mo-0.14N. The material was processed by forging and hot rolling to thick plate then solution treating at 1250°C before being given a 50% cold rolling reduction. On subsequent

30

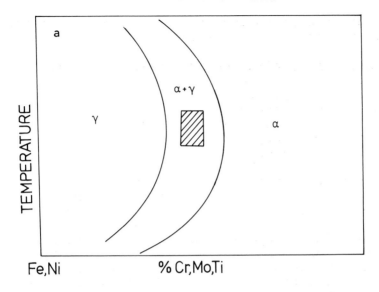

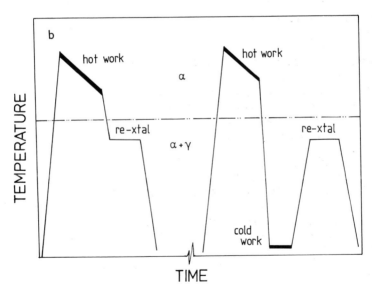

Figure 2.12 (a) Schematic (Fe,Ni)-(Cr-Mo-Ti) phase diagram showing the region where superplasticity is observed. (b) Thermomechanical processing schedules which can be used to develop a microduplex microstructure in Fe-Ni-Cr-Mo alloys.

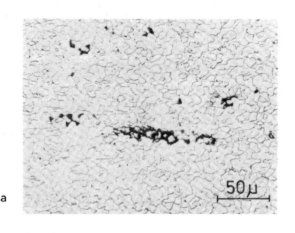

a

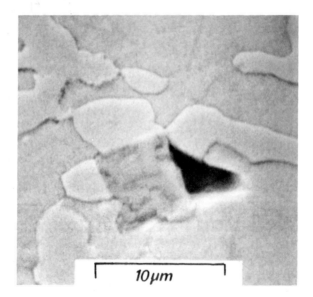

b

Figure 2.13 Cavity formation adjacent to banded
carbide inclusions in IN-744 [66].

deformation at an optimum initial strain rate of
4 x 10^{-3}/s two peaks of high superplastic
elongation (~2500 and ~2000%) were observed at
900°C and 1050°C, respectively. Detailed studies
showed that the high elongations were due to the
evolution of a fine equiaxed microstructure of
$\sigma + \tau$ at 900°C and of $\alpha + \tau$ at 1050°C.

Cavitation is observed to develop during the
superplastic deformation of duplex stainless
steels. In some materials cavities are mainly
associated with Ti(C,N) inclusions and are believed
to have nucleated during the original thermomechanical processing applied to the material in
order to develop the fine grain microstructure [66]
(Fig. 2.13). The small additions of titanium to
materials such as IN744 combine with the residual
carbon to avoid the formation of $Cr_{23}C_6$
precipitates on interphase boundaries which would
normally lead to cracking during hot working.
Cavitation in SAF2205 has been reported to be
mainly associated with triple junctions involving
alpha, gamma and sigma phases, while in 3RE60 the
cavitation behaviour is similar both above and
below the temperatures at which sigma phase forms.

The flow stresses of the particle stabilised
iron based alloys under optimum superplastic
conditions tend to be much higher than those of the
aluminium alloys described in Section 2.5. The
higher stresses make it impractical to bulge form
parts from such alloys since the gas pressures
required would be too high to handle on an
industrial scale. Such alloys are much more
suitable for superplastic isothermal forging which
allows complex shapes to be formed in materials
which are both strong and tough in their as-formed
condition. The flow stresses of the microduplex
stainless steels under optimum deformation
conditions are comparable with those of the
aluminium alloys. However, the high temperatures
required for superplastic forming coupled with the
inherently good cold formability of these
materials may limit the application of
superplasticity in stainless steels.

Table 2.3 Summary of the conditions under which superplasticity is observed in titanium-base alloys (Upper section - α/β alloys: Lower section - fully β alloys)

Alloy	Temperature °C	Strain rate %/min	m	Elongation %	Ref.
Ti-6Al-4V	790 - 940	0.6 - 6	0.8	700 - 1400	18,29,74 76,79,80
Ti-6Al-5V	850	5	0.7	>700	75
Ti-6Al-2Sn-4Zr-2Mo	900	0.6 - 6	0.6 - 0.7	>500	76
Ti-4.5Al-5Mo-1Cr	840 - 870	1.2	0.8	>500	75
Ti-6Al-4V-2Co	815 - 950	0.6 - 6	0.75 - 0.85	720	78,81
Ti-6Al-4V-2Ni	815 - 950	0.6 - 6	0.7 - 0.95	670	78,81
Ti-6Al-4V-2Fe	815	1.2	0.54	650	78
Ti-5Al-2.5Sn	900 - 1100	1.2	0.5	>400	29
Ti-4Al-4Mo-2Sn-0.5Si	880 - 930	0.6 - 6	0.48 - 0.65	500 - 1200	82,83
Ti-15V-3Cr-3Sn-3Al	760 - 850	1.2 - 6	0.5	200	77
Ti-13Cr-11V-3Al	800	-	-	150	77
Ti-8Mn	750	-	0.43	150	84,262
Ti-15Mo	800	-	0.6	100	84

2.7 Titanium alloys

It is somewhat fortuitous that (conventionally processed) alloys such as Ti-6Al-4V and Ti-6Al-2Sn-4Zr-2Mo when hot rolled to sheet can show exceptional superplasticity during deformation in the $\alpha + \beta$ phase field [18,29,74-78]. Since the first report by Lee and Backofen [29] superplasticity has been observed in a range of microduplex titanium alloys and to a lesser extent in fully β titanium alloys (see Table 2.3).

2.7.1 Microduplex alloys

Superplasticity in titanium based alloys is dependent on a number of factors, including the grain size, its distribution and aspect ratio, the relative proportions of the α and β phases, the rate of grain growth and to a lesser extent the texture and inhomogeneity of the starting material. Such a large body of data reflects the extent of commercial interest in superplasticity in titanium alloys. Increasing the grain size results in a shift in the position of Region II (the superplastic regime) to lower strain rates and an increase in the flow stress [18,29]. Strain enhanced grain growth is commonly observed in Ti-6Al-4V and leads to an increase in the flow stress and a reduction in the strain rate sensitivity with increasing strain [18] (Fig. 2.14).

The α and β phases have very different deformation characteristics. The cph α phase has fewer slip systems and a self-diffusivity which is approximately two orders of magnitude slower than that in the bcc β phase. It might therefore be expected that the greater the volume fraction of β phase, the easier it would be to relax the stresses generated during grain boundary sliding, and thus the greater the extent of superplasticity. However, the rate of grain growth in alloys containing a large volume fraction of β phase is high and a substantial volume fraction of the harder, less accommodating α phase is required to minimise grain growth. The effect of variations in the volume fraction β phase on the elongation to

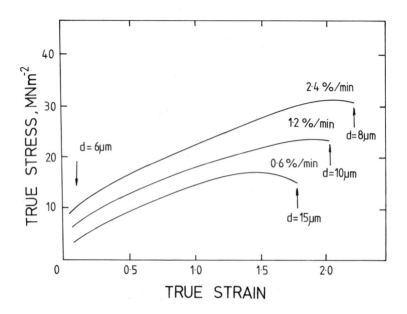

Figure 2.14 Effect of strain enhanced grain growth on the variation of flow stress with strain for Ti-6Al-4V at 920°C (note: grain growth is greatest at the lowest strain rate) [79]

failure and strain rate sensitivity is shown in figure 2.15 [76].

One modification to superplastic titanium based alloys has been the addition of low levels (up to 2 wt-%) of transition metals, such as Fe, Ni and Co, that have high tracer diffusion coefficients [53,54]. This has enabled the temperature at which superplastic deformation is carried out to be reduced by ~100°C from around 915°C to 815°C. The reduction in forming temperature has been possible for two reasons. Firstly, the additions stabilise the β phase allowing the optimum volume fraction of ~40% to be attained at a lower temperature and secondly, the additions increase the self-diffusion

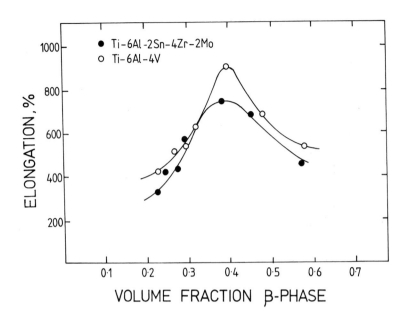

Figure 2.15 Effect of the volume fraction of
β phase on the elongation to failure of two α/β
titanium alloys [76]

rate enabling the normal drop in self-diffusion
rates with falling temperature to be countered.
The lower forming temperatures also enable
reductions in energy costs and tool wear to be
made. Furthermore, grain growth is less extensive
at the lower temperatures and the extent of
strain hardening is significantly reduced.

2.7.2 Fully β alloys

The fully β alloys, after conventional processing,
have a large grain size and the tensile elongations
observed, although high by normal standards (>100%)
are low when compared with those shown by the α + β
alloys [75,77,84]. Examination of the stress-
strain relationship of alloys such as

Ti-15V-3Cr-3Sn-3Al over the normal range of superplastic strain rates (10^{-4} - 10^{-3}/s) initially shows a very high flow stress but after only 1-2% strain this markedly decreases and thereafter remains constant. The microstructural changes that occur during the softening process include the development of a subgrain structure. In effect, the β titanium alloys achieve the high elongations as a result of dislocation and diffusional mass transfer on the scale of the subgrain size rather than the grain size.

2.8 Superplasticity in ceramics

Superplasticity is not normally associated with fully crystalline ceramic materials although it should be recalled that hot glass (an amorphous structure) is the ideal superplastic material, with a strain rate sensitivity of unity and a potential for virtually infinite deformation in either tension or compression. However, with the advent of ultra-fine ceramic powders, the mechanical and microstructural characteristics of superplasticity which are readily observed in a wide range of fine grain metallic materials have also been observed in ceramics. Like their metallic counterparts, two classes of superplastic ceramics exist: single phase and microduplex.

The single phase ceramics are normally prepared from high purity powders which have been milled, sometimes combined with 1-2 wt-% of another ceramic to act as a sintering aid, then hot pressed to give a fully dense material. The majority of the microstructural and mechanical observations of superplasticity in ceramics have been made on specimens deformed in uniaxial compression and are far from comprehensive. However, tensile studies of polycrystalline yttria stabilised zirconia (Y-TZP) [85] have demonstrated that elongations of 170% are easily attainable at commercially viable strain rates (1×10^{-4}/s) with flow stresses less than 10 N mm^{-2}. It would appear that the principle limitation on superplasticity in single phase ceramics is strain enhanced grain growth since there are no second phase particles present within the microstructure to pin the grain boundaries.

Duplex ceramics can be divided into two subgroups: firstly, those which contain a low volume fraction of residual glass such as that resulting from the partial crystallisation of a 'glass ceramic', or from the addition of sintering aids, and secondly, the fully crystalline materials comprising equal volume fractions of two chemically distinct phases (e.g. ZrO_2 with 20 wt-% Al_2O_3). Systematic studies of superplasticity in microduplex ceramic materials are even more limited than for single phase ceramics, despite the potential attractiveness of such materials with regard to segregation stabilisation of the microstructure.

Investigations of high strain deformation in duplex ceramics have been carried out using crystallised β spodumene (lithium aluminosilicate glass) [13] and ZrO_2-Al_2O_3 composites [14,86]. In both instances strain rate sensitivities of 0.5 and greater were reported at strain rates between 10^{-4} and 10^{-3}/s and for the case of β-spodumene tensile elongations up to 400% were observed. The strain attainable in compression was, in both materials, virtually infinite. Strain enhanced grain growth during superplastic flow was still found to occur in the duplex materials but was much less than in the single phase ceramics with comparable initial grain sizes.

The superplastic properties of a number of single phase and microduplex ceramics are summarised in Table 2.4.

2.9 Superplasticity in Other Materials

2.9.1 Nickel alloys

A number of nickel based alloys have been shown to be superplastic when thermomechanically processed to a fine grain microstructure [92]. The microstructures of the nickel alloys are similar to the true microduplex alloys, with a volume fraction of the second phase between 30 and 50%.

Table 2.4 Summary of the conditions under which superplastic behaviour is observed in ceramic materials (Upper section – single phase: Lower section – microduplex)

Alloy	Grain size range, μm	Temperature °C	Strain rate %/min	m	Elongation %	Ref.
UO_2	2 – 6.3	1100–1400	0.02 – 0.1	0.5 – 1	– 68	15
MgO	4	1100–1150	0.06	0.5	– 78	87
Al_2O_3	0.1 – 1	1100–1150	0.04 – 0.4	0.8	– 120	88
	1.6 – 3	1400	–	–	– 95	89
	1.5 – 2.5	1375–1450	0.006 – 0.36	0.65 – 0.7	–	90
$MgO.2Al_2O_3$	4.6	1500–1610	0.4 – 3.6	0.45 – 0.55	– 160	91
ZrO_2	0.3	1450	0.6 – 3	0.5	+ 170	85
$Li_2O-Al_2O_3-6SiO_2$	0.9 – 2.0	1100–1150	0.6 – 6	0.4 – 1	+ 400	13
$ZrO_2-Al_2O_3$	0.8 – 1.6	1400–1550	0.6 – 12	0.5 – 0.8	– 500	85.86
$ZrO_2-Y_2O_3-Al_2O_3$	0.5	1400–1500	0.6 – 3	0.5	+200	14

Two types of alloys exist, firstly those produced by conventional casting and working and secondly, those which are fabricated via the powder route.

The microduplex structure of the conventional nickel-base alloys is produced by first hot working in the single phase region to break down the initial cast structure. Hot working is continued as the temperature is allowed to fall into the the two phase field. Nucleation of the second phase during cooling would normally occur virtually homogeneously throughout the microstructure. Since hot working can continue below the recrystallisation temperature a short anneal high in the two phase field prior to cooling gives rise to an equiaxed microduplex structure (Fig. 2.16). If the kinetics of precipitation of the second phase during hot working or annealing are slow then the material is given an additional cold working treatment prior to recrystallisation within the two phase field to generate a duplex structure. [93,94]. Depending on the alloy composition, the microstructure would consist of equiaxed grains of nickel solid solution,which is strengthened by Co, W or Mo, and a second phase which could be either α-Cr [95] or, if the alloy contains Al, Ni_3Al [96,97]. The average grain size of the processed material would typically be in the range 2 to 7 μm. The range of temperatures and strain rates over which the Ni-base alloys are superplastic are summarised in Table 2.5.

The second group of nickel based alloys are those produced by the powder route. These include materials such as IN-100 [98-100], IN-713 [101] and MAR M-6000 [102]. The alloys are produced from compacted and hot extruded powders. After extrusion, the grains are generally equiaxed and grain size stabilisation is provided by either a high volume fraction of Ni_3Al as in the conventional alloys, by fine grain boundary carbides, or by a fine dispersion of oxide particles, usually yttria (Y_2O_3). The oxide particle size is generally between 25 and 35 nm.

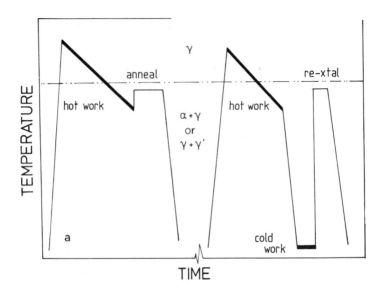

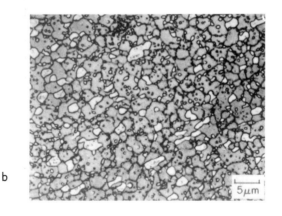

Figure 2.16 (a) Schematic processing schedule used
to develop fine grain microstructures in duplex
Ni-base alloys. (b) Microstructure of superplastic
Ni-Cr-Fe alloy in the recrystallised condition (the
lighter phase is α-Cr) [95].

Table 2.5 Summary of the superplastic properties of some nickel-based alloys

Alloy	Temperature °C	Strain rate %/min	m	Elongation %	Ref.
Ni-10Cr-15Co-5Al-5Ti-3Mo (IN-100)	920-1050	2 - 50	0.5 - 0.65	>850	100
Ni-16Cr-8Co-3W-3Al-3Ti (IN-738)	900	0.6 - 6	0.4	500	97
Ni-39Cr-8Fe-2Ti-2Al	980	0.6 - 60	0.5	>1000	95
Ni-13W-11Co-10Cr-5Al-5Ti (MAR M-200)	1050-1180	0.6 - 6	0.58	-	103
Ni-12Cr-6Al-5Mo-2Nb-1Ti	1050	0.6 - 6	0.5	-	101
Ni-8Cr-9Co-9W-5Al-3Ta-1Hf (MAR M-247)	1050	-	0.63	>800	104
Ni-15Cr-5Al-4W-2Ti-2Ta-2Mo -1.1 Y₂O₃	900-1000	60	-	150 - 310	102
8(Ni₃Si)-17(Ni₃B)	550	0.6	0.5	120	261

43

The high flow stresses of the superplastic Ni-base alloys tend to favour their use as isothermal forging alloys with their high inherent ductility allowing complex shapes to be simply fabricated. For example, IN-100 powder source alloys are forged to produce turbine discs with integral blades. Following forging, the blades on the outer circumference are selectively heat treated to improve their creep resistance under service conditions [97].

Superplasticity has also been developed in MAR M-247 [104] by thermomechanical processing of compacts which had been extruded from bundles of rapidly solidified ribbons.

2.9.2 Copper alloys

Superplastic behaviour has been reported for a range of α/β copper alloys including brasses, aluminium bronzes and nickel-silvers, and also for the essentially single phase copper dispersion alloy, Coronze (CDA-638). For the binary brasses and aluminium bronzes the duplex superplastic microstructure is obtained by hot working the materials in the $\alpha + \beta$ phase field or during cooling from the β to the $\alpha + \beta$ fields.

The optimum deformation temperature for the brasses is ~600°C with maximum tensile elongations of ~500% being obtained for an alloy with approximately equal volume fractions of the two phases. However, the alloys cavitate and since accommodation of superplastic flow is believed to occur primarily within the β phase, the level of cavitation increases as the volume fraction of α phase increases. For the aluminium bronzes the optimum deformation temperature is ~700°C. Although the binary α/β copper alloys would be classified as segregation stabilised, the composition difference between the two phases is relatively small (4 to 5%) so that grain growth is not strongly inhibited.

The α/β nickel silvers such as IN-836 and IN-629, see Table 2.6, are processed by hot working in the α phase field, solution treating and then quenching

44

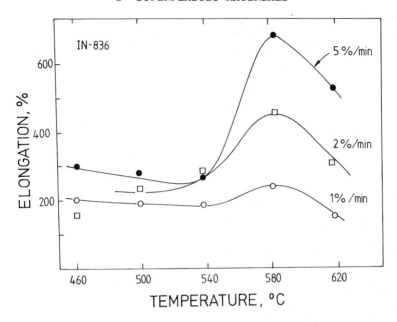

Figure 2.17 Variation in the elongation to failure with temperature of IN-836 [70]

to room temperature to retain the single phase structure. After a heavy cold rolling reduction, ~85%, the material is finally annealed in the two phase field to recrystallise the α phase with a grain size of 1-2 μm and precipitate a dispersion of fine, d ~1 μm, β phase particles. The optimum deformation temperatures are 600°C and 570°C (Fig. 2.17) respectively, but the alloys are prone to cavitation due to the low volume fraction of β phase in the alloy and the relatively low deformation temperatures [105-108].

The dispersion strengthened copper alloy CDA-638 is a cobalt modified Si-Al bronze. It is a pseudo single phase alloy in which a fine α grain size of ~1 μm is produced by recrystallisation following heavy cold deformation. The grain size is stabilised by a fine dispersion of cobalt silicide particles and the alloy is superplastic at

~500°C, but is particularly prone to cavitation
[109-112].

Exceptional superplasticity has been observed in a
complex Al-bronze of nominal composition
Cu-10Al-5Fe-5Ni. A fine stable microstructure can
be developed by a processing sequence which
involves rolling at 900°C, annealing at 700°C,
warm rolling at 640°C, cold rolling and finally
recrystallisation at ~800°C. The room temperature
microstructure consists of α phase with a grain
size of 2-3 μm containing a fine dispersion of
aluminide particles (K phases). The material can be
superplastically deformed at 750-850°C which spans
the α + K and the α + β + K phase fields. Tensile
elongations of 1000% can be readily obtained at
strain rates as high as 10^{-2} to 10^{-1}/s, with an
elongation of 5,500% without failure being recorded
for an initial strain rate of 6 x 10^{-3}/s at 800°C.
At this temperature, the microstructure contains
about 40% α, 30% β and 30% K by volume and its
stability is due to a combination of segregation
and particle stabilisation.

2.9.3 Eutectics and eutectoids

Superplasticity has been reported in over 100
different alloys. It is beyond the scope of this
book to detail all the recorded instances of
superplasticity and the reader is referred to the
review of Padmanabhan and Davies [10] for further
details. However, the properties of four eutectic
alloys (Mg-33Al, Pb-62Sn, Bi-44Sn and Ag-28Cu)
together with those of a commercially used eutectoid
alloy (Zn-22Al) are given in Table 2.7.

Fine grain microstructures are formed in the
eutectic alloys by recrystallisation following
extrusion. In the case of the Zn-22Al alloy, a fine
grain microstructure is formed by the spinodal
decomposition of a Zn-rich solid solution quenched
from above the monotectoid temperature.

Table 2.6 Summary of the superplastic properties of copper-base alloys.

Alloy	Temperature °C	Strain rate %/min	m	Elongation %	Ref.
Cu-38Zn-15Ni-0.2Mn (IN-836)	460 - 620	0.01 - 4	0.37 - 0.5	200 - 680	105,107
Cu-28Zn-15Ni-13Mn (IN-629)	570	0.6 - 3	0.5	440	106,108
Cu-4Al-2Si-0.4Co (CDA-638)	470 - 600	0.01 - 1	0.35 - 0.46	200 - 320	109-112
Cu-10Al-5Fe-5Ni	750 - 850	30 - 600	-	1000 - 5500	2

Table 2.7. Summary of the conditions under which superplasticity is observed in eutectic/eutectoid alloys.

Alloy	Temperature °C	Strain rate %/min	m	Elongation %	Ref.
Mg-33Al	350 - 400	0.6 - 60	0.6 - 0.8	- 2100	16
Pb-62Sn	20 -100	-	0.5 - 0.7	- 4850	1,6,113
					114
Bi-44Sn	30	-	-	1950	6
Zn-22Al	200 - 250	0.1 - 10	0.4 - 0.66	500 - 2900	9,38,115 116,117
Ag-28Cu	650 - 730	0.025	-	500	118

3

CHARACTERISATION OF SUPERPLASTIC DEFORMATION BEHAVIOUR

3.1 Introduction

In this chapter, the experimental procedures which can be adopted to measure the extent to which a material may be deformed superplastically are described. A number of worked examples are included in the text to guide the reader through the various interpretative procedures.

As was stated in the introductory chapter, the most important mechanical characteristic of superplasticity is the high strain rate sensitivity of the flow stress, m, and its variation with strain rate, temperature and grain size. In an ideal material, where the microstructure remains constant, the true flow stress can be measured by carrying out constant strain rate tensile tests at a range of strain rates and measuring the steady state load. However, because of the experimental difficulties in obtaining low strain rates in tensile testing equipment, constant stress creep testing can be more usefully employed to determine the steady state strain rate at the lower stresses. From eqn. (1.1)

$$\log (\sigma) = m \log (\dot{\epsilon}) + k'' \qquad (3.1)$$

The slope of a logarithmic plot of the true stress against true strain rate gives the strain rate sensitivity, m.

Most of the stress-strain rate and stress-strain data presented in the literature has been obtained from constant crosshead velocity tensile testing and cannot be regarded as truly steady state. During constant velocity testing, the true strain rate decreases as the specimen elongates and thus the flow stress, if all other effects remain unaltered, would decrease with increasing strain. This situation is the reverse of that found in constant load creep testing where the specimen cross-sectional area decreases with increasing strain. The test piece thereby experiences a continuously increasing stress during testing. However, the problem in superplasticity is much more pronounced than in creep as uniform elongations of hundreds of percent are common and it is therefore possible for the strain rate to decrease by more than an order of magnitude during constant velocity testing.

Most engineering materials are microstructurally unstable at elevated temperatures. Hence, it is impossible to determine a meaningful flow stress from a constant strain rate tensile test, since the flow stress often increases with increasing strain due to the effects of grain growth. It is therefore important, where comparisons are to be made, to determine the flow stress at a constant structure. To do this, step strain rate or strain rate jump tests have been developed.

3.2 Step strain rate tests

To determine the strain rate range over which a material might be expected to exhibit superplastic behaviour, a series of step strain rate tests are carried out at various temperatures. Strain rate or velocity jump tests were originally developed by Backofen et al. [9] for the determination of strain rate sensitivity. Initially a tensile specimen is deformed at a constant velocity until a steady load is registered. As many materials do not have a uniform starting structure, a strain of 30 to 50% is usually allowed to accumulate before the first velocity jump is made. After the velocity jump, which can be either upwards or downwards, a new

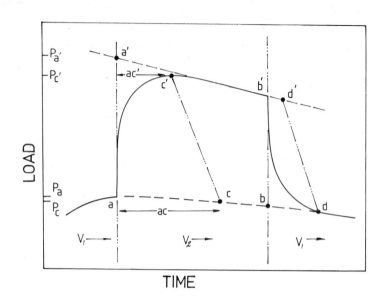

Figure 3.1 A schematic load – time plot for a velocity jump test on Supral 220 at 460°C (V_1=0.02 mm/min; V_2=0.05 mm/min)

steady state is allowed to develop before returning the crosshead velocity to its original value. A record of load against time for a velocity jump test carried out on Supral 220 is shown in figure 3.1.

Several procedures have been proposed for determining the relevant true stresses and strain rates at each crosshead speed. Backofen proposed that following the velocity increase, the maximum load, Pc' should be measured. The load versus time plot at the initial testing velocity should be extrapolated to the strain at which Pc' was measured and the corresponding load, Pc, determined. A similar procedure can be adopted for determining the respective loads for a velocity

50

decrease. Provided that the velocity jump is small then the strain rate sensitivity can be assumed to remain constant over the range of strain rates experienced by the test piece and

$$m = \frac{\log(Pc'/Pc)}{\log(V_2/V_1)} = \frac{\log(Pd/Pd')}{\log(V_1/V_2)} \qquad (3.2)$$

for the velocity increase and decrease, respectively.

Using figure 3.1, which is a diagrammatic representation of the load-time plot and is not to scale, an example of the procedure used to determine 'm' would be as follows:

The velocity of the crosshead of the tensile test machine is increased from 0.02 mm/min to 0.05 mm/min and the load allowed to stabilise before returning the crosshead to its original speed. The load-time record shown in figure 3.1 has been recorded with a chart speed of 20 mm/min. The load-time plot at the initial velocity is first extrapolated past the point at which the velocity jump was made, to rejoin the trace where a second steady state load has redeveloped at the original testing velocity. The point of maximum load, c', at the increased velocity is located, and the load Pc', noted. (In this example the maximum load was 5.3 kg).

The time, t_2, taken to attain the maximum load from the point at which the velocity was changed, a, is calculated from the distance between the point c' and a.

$$t_2 = \frac{\text{distance } (a-c')}{\text{chart speed}} = \frac{28 \text{ mm}}{20 \text{ mm/min}} = 1.4 \text{ min} \qquad (3.3)$$

Next, the time, t_1, that would have been required at the initial testing velocity to reach the strain corresponding to point c', is calculated.

$$t_1 = t_2 \times \frac{V_2}{V_1} = 1.4 \text{ min} \times 0.05/0.02 = 3.5 \text{ min} \qquad (3.4)$$

The corresponding distance along the extrapolated line a-d would then be the product of the chart speed and the time i.e. 20 mm/min x 3.5 min = 70 mm. The load at this point, Pc, is then measured. In this case the load was found to be 3.8 kg.

Using eqn (3.2)

$$m = \frac{\log(Pc'/Pc)}{\log(V_2/V_1)} = \frac{\log(5.3/3.8)}{\log(0.05/0.02)} = 0.36 \quad (3.5)$$

The corresponding value obtained for the velocity decrease is m = 0.36. Alternatively the load-time plot at the increased testing velocity can be extrapolated back to the point at which the velocity change was made and the corresponding loads measured [11]. Then from figure 3.1,

$$m = \frac{\log(Pa'/Pa)}{\log(V_2/V_1)} = \frac{\log(5.5/3.9)}{\log(0.05/0.02)} = 0.38 \quad (3.6)$$

A similar procedure can be adopted if the tests are carried out at a constant true strain rate rather than a constant velocity [21].

To measure m over a range of strain rates, a tensile specimen is deformed at a velocity which will produce a strain rate in the middle of the range of interest until a steady state is attained. The crosshead speed is then reduced to a low value and the load allowed to stabilise before being measured. Repeated incremental increases in crosshead velocity allow the load to be measured at a range of strain rates [119]. A typical load-time plot is shown in figure 3.2. The true stresses and corresponding true strain rates are then calculated from the maximum loads, crosshead velocities and instantaneous lengths of the sample. The derivative of the best fit curve to a logarithmic plot of true stress against true strain rate then gives the strain rate sensitivity, m, and its variation with strain rate in the range of interest.

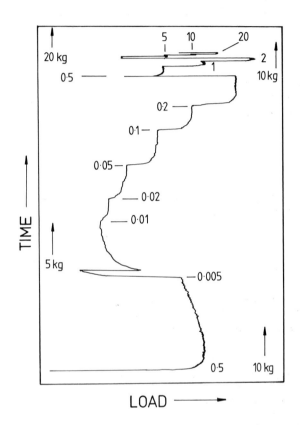

Figure 3.2 A record of load vs. time for a repeated velocity jump test carried out on Al-8090 at 500°C (figures indicate the crosshead speed in mm/min)

Table 3.1 Tabulated record of the velocity jump test shown in figure 3.2 illustrating the procedure used to calculate the true stress and true strain rate.

Crosshead speed, v_i (mm/min)	Distance d_i (mm)	Specimen length, l_i (mm)	Strain rate $\dot{\epsilon}_i$ (10^{-6}/s)	Area a_i (mm²)	Load P_i (kg)	Stress σ_i (Nmm⁻²)	$\log_{10}(\sigma_i)$	$\log_{10}(\epsilon_i)$
0.5	80	10	595	10.089	5.3	7.21	0.85	-3.23
0.005	47	14.0	5.94	7.206	1.20	1.64	0.21	-5.22
0.01	17	14.0235	11.9	7.194	1.33	1.82	0.26	-4.93
0.02	28	14.0405	23.6	7.186	1.65	2.26	0.35	-4.63
0.05	30	14.0965	58.5	7.157	2.20	3.05	0.48	-4.23
0.1	20	14.247	115	7.082	2.83	3.97	0.60	-3.94
0.2	21	14.447	224	6.984	3.65	5.28	0.72	-3.65
0.5	12	14.867	539	6.786	4.65	6.99	0.84	-3.27
1.0	5	15.47	1040	6.523	6.10	9.47	0.98	-2.98
2.0	3	15.97	2010	6.319	8.00	12.9	1.10	-2.70
5.0	2	16.57	4740	6.089	9.6	16.4	1.21	-2.32
10	0.5	17.57	9220	5.742	11.7	20.56	1.31	-2.04
20	0.5	18.07	17500	5.583	13	24.2	1.38	-1.76
		19.1		5.28				

Chart speed   = 10 mm/min
Specimen gauge length   = 10 mm
Specimen thickness   = 2.03 mm
Specimen width   = 4.97 mm

The procedure used to determine the strain rate sensitivity from a series of velocity jump experiments is explained in the following section.

3.3 Determination of strain rate sensitivity

The chart shown in figure 3.2 is a record of load against time (equivalent to displacement) for a superplastic aluminium alloy tested using the method outlined above. To determine the strain rate sensitivity and its variation with strain rate, the true flow stress and instantaneous strain rate need to be determined at each crosshead velocity. The points corresponding to the velocity changes are first marked on the chart and the instantaneous length of the sample, l_i, is calculated using

$$l_i = l_{i-1} + v_i d_i/\text{chart speeed} \qquad (3.7)$$

where v_i is the crosshead velocity immediately prior to the velocity change, d_i is the distance on the chart covered at that velocity and l_{i-1} is the length of the test piece at the previous speed change.

The initial gauge length of the test piece used to obtain figure 3.2 was 10 mm and the sample was strained to an elongation of 40% prior to the first velocity change. Using figure 3.2 as an example then the length of the sample at the initial velocity change would be 14 mm and at the second velocity change 14.0235 mm (= $14 + 47 \times 0.005/10$), see Table 3.1. Thus for the third velocity change

$$l_3 = 14.0235 + 17 \times 0.01/10 = 14.0405 \text{ mm} \qquad (3.7a)$$

The instantaneous strain rate, $\dot{\epsilon}_i$, is then given by

$$\dot{\epsilon}_i = v_i/l_i \qquad (3.8)$$

which for the above example would be

$$\dot{\epsilon}_3 = (0.01/60)/14.0405 = 1.19 \times 10^{-5}/s \qquad (3.8a)$$

To calculate the true flow stress, the steady state load, P_i, at that velocity is measured from

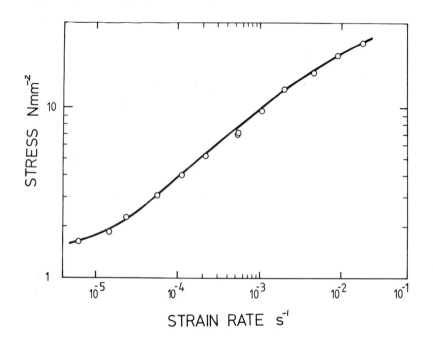

Figure 3.3 The variation of true stress with true strain rate (plotted logarithmically) as derived from the 'm' test record given in figure 3.2.

the chart and the cross-sectional area, a_i, calculated from the instantaneous length of the sample assuming both constant volume and a uniform cross-sectional area, i.e.

$$a_i = l_o a_o / l_i \qquad (3.9)$$

where l_o and a_o are the original gauge length and cross-sectional area, respectively. Continuing the example,

$$a_3 = \frac{10.0 \times (4.97 \times 2.03)}{14.0405} = 7.186 \text{ mm}^2 \qquad (3.9a)$$

The true stress, σ_i, is then calculated from

56

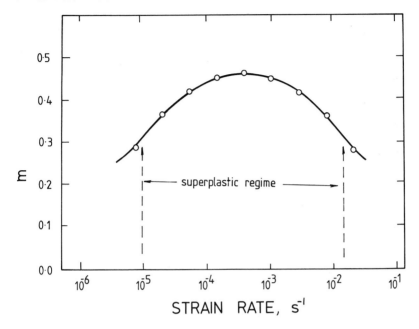

Figure 3.4 The variation of strain rate
sensitivity with strain rate as calculated from the
polynomial expression defining the best-fit curve
in figure 3.3.

$$\sigma_i = P_i / a_i \qquad\qquad (3.10)$$

hence

$$\sigma_3 = \frac{1.33 \text{ kg} \times 9.81 \text{ m s}^{-2}}{7.186 \text{ mm}^2} = 1.816 \text{ N mm}^{-2} \qquad (3.10a)$$

The data obtained from the load versus time record
shown in figure 3.2 is summarised in Table 3.1
together with the calculated values of stress and
strain rate. A logarithmic plot of true stress
against true strain rate (the last two columns of
Table 3.1) is shown in figure 3.3 together with

the best fit curve through the data points. The strain rate sensitivity at a given strain rate can be calculated either by constructing tangents to the curve or by evaluating the derivative of the polynomial expression defining the curve. The calculated variation of 'm' with strain rate is plotted in figure 3.4.

3.4 Determination of the true stress as a function of strain for deformation at a constant strain rate

Conventional tensile tests are carried out at constant crosshead velocity. The large tensile strains which can be accumulated in superplastic materials result in a considerable reduction in the true strain rate during testing. A decrease in the true strain rate would, because of the high strain rate sensitivity of superplastic materials, normally produce a marked decrease in the true flow stress. To allow a steady state deformation stress to be measured it is important to maintain a constant true strain rate during tensile testing. The actual experimental method adopted to produce a true strain rate will depend on the make of tensile testing frame available.

A record of load against time for a superplastic aluminium alloy deformed at a constant true strain rate of 7.8×10^{-4}/s is shown in figure 3.5. The strain at any point on the chart is given by

$$\epsilon = \dot{\epsilon} t$$

$$\epsilon = \dot{\epsilon} \times (\text{distance on chart/chart speed}) \qquad (3.11)$$

Thus for the point marked A in figure 3.5 (chart speed = 10 mm/min), the true strain is given by

$$\epsilon = 7.8 \times 10^{-4} \times \frac{130 \text{ mm}}{(10/60 \text{ mm/s})} = 0.6084 \qquad (3.11a)$$

The instantaneous cross-sectional area is

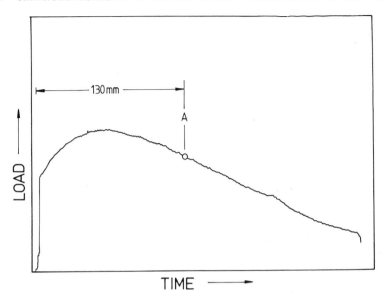

Figure 3.5 A load-time record for a constant strain rate tensile test carried out on Al-8090 ($\dot{\epsilon}$=7.8x10^{-4}/s; 520°C)

calculated by assuming that the strain is uniform i.e.

$$a = a_o \exp(-\epsilon) \qquad (3.12)$$

Hence,

$$a = (\text{width x thickness}) \exp(-\epsilon)$$

$$a = (5.02 \text{ mm x } 1.98 \text{ mm}) \exp(-0.6084) \doteq 5.409 \text{ mm}^2 \quad (3.12a)$$

The true stress, σ, is then given by eqn (3.10) and so

$$\sigma = \frac{4.1 \text{ kg x } 9.81 \text{ m s}^{-2}}{5.409 \text{ mm}^2} = 7.44 \text{ N mm}^{-2} \qquad (3.13)$$

59

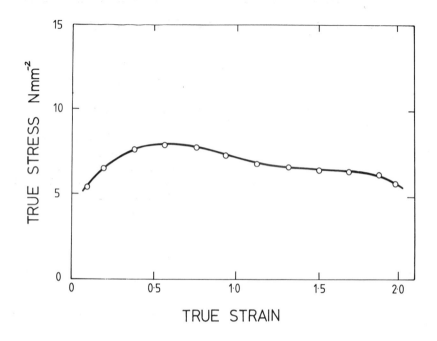

Figure 3.6 The variation of true stress with true strain derived from figure 3.5.

The variation of true stress with true strain derived from figure 3.5 is shown in figure 3.6 and it is apparent that the flow stress is almost independent of strain. In an ideal superplastic material the flow stress is independent of strain. The variation of stress with strain can give an insight into the microstructural changes which occur during superplastic deformation. An increase in the flow stress with strain is normally indicative of strain enhanced grain growth while a decrease in the flow stress, particularly at high strains can often infer the development of cavitation damage (see chapter 5).

3.5 Determination of an activation energy for superplastic flow

In order that the activation energy for superplastic flow, Q_{sp}, can be determined, the temperature dependence of the relationship between flow stress and strain rate must be known on, at least, an empirical level. (The detailed form of such an equation will be discussed in Chapter 4). Following the conventional phenomenological analyses of high temperature deformation, the strain rate is assumed to depend on temperature at a given stress rather than the stress being a function of temperature at a given strain rate. To a first approximation

$$\dot{\varepsilon} = A \frac{\exp(-Q_{sp}/RT)}{kT} \sigma^{1/m} \qquad (3.14)$$

which (if the temperature range, dT, is narrow) can be rewritten as

$$\ln(\dot{\varepsilon}) = B + 1/m \ln(\sigma) - Q_{sp}/RT \qquad (3.15)$$

where A and hence B are material constants. Therefore, providing that the strain rate sensitivity remains constant over a range of temperatures then the activation energy is simply obtained from the slope of an Arrhenius plot of ln (strain rate) against the reciprocal of the absolute temperature at constant stress. Unfortunately, a number of errors can be readily introduced into the determination of activation energy for superplastic flow. These can be summarised as follows:

1. In practice the temperature interval, δT, is not insignificant, usually 30°C or more.
2. The strain rate sensitivity, m, varies with temperature at a given stress.
3. The constant A, and hence B, contains a number of temperature dependent terms.
4. Eqn (3.14) assumes a constant structure. In many duplex materials the phase proportions change

dramatically with small differences in
temperature.
5. It is often argued that the measurements should
   be made for a constant value of the modulus
   compensated stress rather than the actual stress.
   The modulus is temperature dependent.
6. The exact form of eqn (3.14) is not known.

The factors noted above often make it difficult to
interpret or assign the measured activation energy
to a particular physical process and in many
instances the measured activation energy is itself
strongly temperature dependent.

The variation of the flow stress with strain rate
measured using a step strain rate test for a
superplastic material is shown schematically in
figure 3.7 for tests conducted over a range of

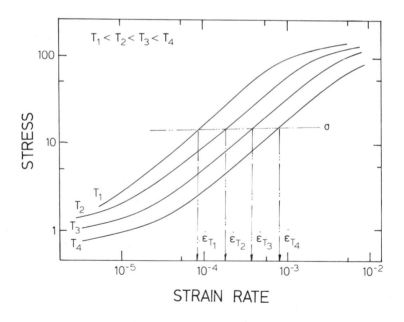

Figure 3.7 Schematic variation of flow stress with
strain rate for a superplastic material.

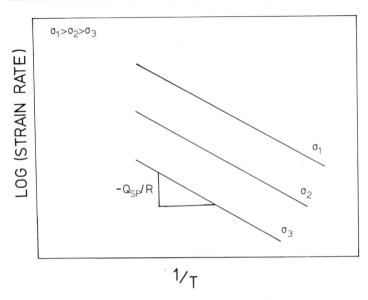

Figure 3.8 Schematic Arrhenius plot used to determine the activation energy for superplastic flow at constant stress.

temperatures. To evaluate the activation energy for superplastic flow, a stress (or several stresses) which lie within the Region II at all temperatures are chosen. For this example, which is based on data obtained for Ti-6Al-4V over the temperature range 790°C to 940°C, a stress of $15 \, N \, mm^{-2}$ was selected. The measured values of the strain rate sensitivity at each test temperature exhibited very little variation and the term $1/m \ln(\sigma)$ is therefore assumed constant. The strain rates corresponding to selected temperatures are next read from the stress-strain rate curves. The strain rates are then plotted as a function of $1/T$ as shown in figure 3.8, for 3 constant stress levels.

It can be seen from eqn.(3.15) that the slope of the Arrhenius plot is equivalent to $-Q_{SP}/R$. The gradients of the lines at each stress are then determined by a least squares analysis. (In the case of Ti-6Al-4V, the calculated slope was found to vary from -2.15×10^4 to -1.99×10^4 as the stress was increased from 10 to 25 N mm^{-2}. The corresponding activation energies for superplastic flow range from 179 kJ/mol at 10 N mm^{-2} to 165 kJ/mol at 25 N mm^{-2}.) If the temperature dependencies of the strain rate sensitivity, shear modulus and the diffusion coefficients included in the constant A are taken into consideration, and the strain rates are measured at a constant value of the modulus compensated stress, then the mean activation energy for superplastic flow is often found to be close to that for grain boundary diffusion in the material being studied. (For both α and β titanium Q_{gb} = 153 kJ/mol.)

3.6 Summary

The experimental and analytical methods which can be used to characterise the mechanical behaviour of superplastic materials have been described. In assessing whether or not a material is superplastic it is first necessary to evaluate the strain rate sensitivity over a range of strain rates at several temperatures in the 'superplastic range'. For microduplex materials the deformation temperatures are ~$0.6T_m$ while for the pseudo single phase alloys the deformation temperatures can often be in excess of $0.8T_m$. Values of the strain rate sensitivity greater than 0.3 are usually indicative of superplastic behaviour. After determining the strain rate sensitivity, constant strain rate tensile tests are carried out at the temperatures and strain rates where m is a maximum. Elongations to failure of 400% and greater are a good indication that true superplasticity has been achieved.

4

PHENOMENOLOGY OF SUPERPLASTIC DEFORMATION

4.1 Introduction

Several plausible hypotheses for the origin of superplasticity in fine grain materials have been suggested yet none has been found capable of accurately describing both the mechanical and microstructural features of superplastic deformation.

The strain rate at which a material will deform at elevated temperature is defined by equation (1.1) which can be restated in an expanded temperature dependent form as:

$$\dot{\epsilon} = A \, \frac{DGb}{kT} \left[\frac{b}{d} \right]^p \left[\frac{\sigma}{G} \right]^n \qquad (4.1)$$

where p and n are characteristic of the micromechanism of deformation, D is a diffusion coefficient which is dependent on the rate controlling process and defines the temperature dependence of the strain rate at constant stress and structure, d is the grain size and b the Burgers vector or characteristic dimension. Deformation is driven by the deviatoric (shear) component of the applied stress field, characterised by σ in eqn (4.1) and G is the shear modulus.

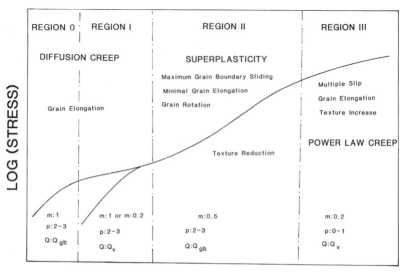

Figure 4.1 Schematic illustration of the strain rate dependence of flow stress in a superplastic material [141].

A logarithmic plot of stress vs strain rate for a typical superplastic material is shown in figure 4.1 and can be divided into 4 regions. The slope of the plot is the strain rate sensitivity, m, where

$$m = \frac{d \log(\sigma)}{d \log(\dot{\epsilon})} = 1/n \qquad (4.2)$$

At high strain rates, i.e. Region III, the strain rate sensitivity is low (m = 0.2 to 0.3) and deformation is by recovery controlled dislocation creep. The activation energy for flow in Region III is similar to that for lattice diffusion and the strain rate is relatively insensitive to the grain size. At intermediate strain rates, Region II, the

66

material deforms superplastically, with grain
boundary sliding being a major feature of the flow
process. The strain rate sensitivity is high
($m \geq 0.5$). The measured activation energy for flow
in Region II is often similar to that for grain
boundary diffusion and the flow stress is grain
size dependent ($p = 2$ to 3).

At low strain rates, Region I, the slope of the
stress-strain rate plot varies. In many
superplastic alloys the strain rate sensitivity in
Region I is low and this has been interpreted as
evidence for some form of threshold stress for
superplastic flow since dislocation activity is not
normally observed at such low strain rates.
However, in other materials the slope at low strain
rates has been observed to increase, tending
toward unity. This has been cited as evidence for
diffusion controlled flow with the activation
energy for flow in Region I being similar to that
measured for volume diffusion. This is contrary to
that which might normally be expected in a fine
grain material, i.e. the activation energy would be
that for grain boundary diffusion. Moreover,
different studies on the Zn-Al eutectoid have
shown both types of Region I behaviour.

To add further to the controversy, it has been
reported that the strain rate sensitivity can
initially decrease as the strain rate is reduced
but further reductions in the strain rate result in
a transition to a slope of unity, termed Region 0,
where true diffusion creep dominates [120].
Unfortunately, it is difficult to resolve the
arguments concerning Region I as the materials
examined were microstructurally unstable. Grain
growth during slow strain rate deformation or
prolonged periods of primary creep have both been
cited as the reasons for the apparent discrepancies
between different sets of experimental data. These
factors coupled to the different testing techniques
that have been used to obtain stress-strain rate
data at very low strain rates have complicated the
interpretation of the low strain rate regime [12].

4.2 Non-superplastic flow - Region III

At high strain rates, i.e. Region III, the strain
rate sensitivity is low and the material is
deforming by conventional recovery controlled
dislocation creep. Strain is accumulated by the
glide of dislocations within the grains.
Dislocation glide is opposed by the microstructure
of the material, for example, on a near atomic
scale by other dislocations, solutes, even the
lattice itself, and on a larger scale by
precipitates and dislocation arrays within the
grains. Glide is therefore dependent on the rate at
which the obstacles can be bypassed. At elevated
temperatures, thermal activation enables
dislocations to either absorb or emit vacancies at
a realistic rate and to climb from one glide plane
to another thereby overcoming the obstacles.
Similarly, if the dislocation is trapped as a
consequence of jogs formed by the intersection with
forest dislocations then the absorption or emission
of vacancies will enable the gliding dislocation to
be released. Those dislocations which become
trapped in the sub-boundaries will climb and
either be annihilated in the boundary or escape.
The strain rate is therefore controlled by the rate
at which dislocations are made available for glide,
the glide time being negligible with respect to the
time spent at pinning points [121].

Several theoretical treatments, each of which
assume that dislocation climb in one form or
another is the rate controlling process have led to
values for the stress exponent, n, ranging from 3
to 5 when volume diffusion is the dominant mode of
vacancy supply. At lower temperatures vacancy
diffusion along the dislocation cores predominates.
The rate at which vacancies are supplied to or
removed from the climbing dislocations is then
dependent on the dislocation density which is
itself a function of the applied stress. Stress
exponents of 5 to 7 (n + 2) are then predicted. As a
result of continued slip on a limited number of
active slip systems, the grains become elongated
parallel to the tensile axis during deformation.
Grain boundary sliding, whilst operative in Region

III, is only a minor flow mechanism [122]. Since the distances over which dislocations move between pinning points are small when compared with the grain size, then the strain rate is virtually independent of the grain size. The phenomenological equations defining this region are discussed in detail elsewhere [123], but can be approximated to

$$\dot{\epsilon}_{III} = A_{III} \frac{D_{eff}Gb}{kT} \left[\frac{\sigma}{G}\right]^n \qquad (4.3)$$

where

$$D_{eff} = D_v + \frac{\pi a_c}{b^2} \left[\frac{\sigma}{G}\right]^2 D_c \qquad (4.4)$$

where A_{III} is a constant, D_{eff} is the effective lattice diffusion coefficient, D_v the volume diffusion coefficient, D_c the dislocation core diffusion rate, a_c the dislocation core cross-sectional area. The presence of a deviatoric component in the applied stress field will result in a difference between the normal stresses on some grain boundary surfaces, introducing a gradient in chemical potential which in turn induces diffusional mass transfer to reduce those differences. Diffusional transport both through and around the grains results in a shape change and hence creep [124,125]. The strain rate for true diffusion creep, Region 0, is given by

$$\dot{\epsilon}_0 = A_0 \frac{D_{diff}Gb}{kT} \left[\frac{b}{d}\right]^2 \left[\frac{\sigma}{G}\right] \qquad (4.5)$$

and

$$D_{diff} = D_v + \frac{\pi \delta}{d} D_{gb} \qquad (4.6)$$

where A_0 is a constant, D_{gb} is the grain boundary diffusion coefficient, δ the grain boundary width, and d the grain size.

At sufficiently low stresses, diffusion creep will dominate over dislocation creep. At low temperatures, the activation energy for diffusion creep will be that for grain boundary diffusion with a grain size exponent $p = 3$, while at elevated temperatures, lattice diffusion will dominate with $p = 2$. If the vacancy creation or absorption processes rather than vacancy motion are rate controlling, as might occur when solutes or precipitates are present in the grain boundaries, then the deformation rate may become interface controlled. The strain rate would then vary as the square of the applied stress [123].

4.3 Superplastic flow - Region II

At intermediate strain rates (Region II) the flow process is less well understood, although there is agreement on the microstructural features associated with it. Strain is accumulated by the motion of individual grains or clusters of grains relative to each other by sliding and rolling. Grains are observed to change their neighbours and to emerge at the free surface from the interior [126]. During deformation the grains remain equiaxed, or, if they were not equiaxed prior to deformation, become so during superplastic flow. Textures become less intense as a result of deformation in Region II, while the converse is normally observed in Region III. The motion of individual grains is dependent on both the normal and shear stresses acting on their grain boundaries and is therefore dependent on the shape and orientation of the grains. Translation and rotation are thus stochastic in nature, occurring in different directions, to different extents, at different locations.

Many attempts have been made to develop theories capable of predicting both the mechanical and topological features of superplastic deformation. However, none have yet been completely successful. Unlike recovery controlled dislocation creep, where dislocation climb is accepted as the rate controlling process, several rate limiting mechanisms have been proposed for superplastic

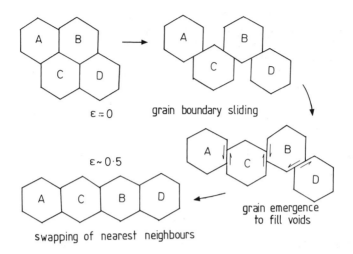

Figure 4.2 Unaccommodated grain boundary sliding. The size of the holes between the grains will be reduced as a consequence of grains moving perpendicular to the plane of motion shown [131,132].

deformation (see reviews by Mukherjee [127,128], Kashyap and Mukherjee [129], Langdon [12,130] and Gifkins [131]).

If grain boundary sliding was to occur in a completely rigid system of grains then voids would develop in the microstructure [132] (Fig. 4.2). The holes or cavities would expand or contract as grains, moving in three dimensions, approached or receded from them. However, many superplastic materials do not cavitate. Grain boundary sliding is therefore accommodated. Even when cavities are observed, their distribution is far from homogeneous and while they would accommodate sliding, cavitation is not as likely an accommodation mechanism as either diffusion or dislocation activity. If the accommodation processes are sufficiently rapid at the deformation temperature, then grain boundary sliding itself could be the rate controlling mechanism [133]. Alternatively, if grain boundary sliding was intrinsically rapid [134,135] then the accommodating processes would be rate limiting.

Microstructural studies have found only limited evidence of dislocation activity within the grains of materials deformed superplastically, although as the strain rate approaches that of the transition to Region III this becomes increasingly apparent [136-141]. By way of contrast, large grained non-superplastic microstructures deforming under the same conditions of strain rate and temperature as those associated with Region II show extensive dislocation activity, often with a well defined substructure present. The dislocation density in the subgrains themselves is, however, generally very low.

The fine grain size of superplastic materials, coupled with low flow stresses at a given strain rate and temperature, ensures that the equilibrium subgrain size is greater than the grain size [142]. It might then be argued that the material is deforming by conventional dislocation creep and that the dislocations are not observed as most of them are trapped within the grain boundaries. However, the absence of dislocations within the grains may also be cited as evidence in support of superplastic flow accommodated solely by diffusion.

In examining the models which have been proposed to explain superplastic flow, it is important to distinguish between cause and effect. Often, it is assumed that grain boundary sliding occurs as a result of the applied stress and that the transient stresses generated as a consequence of the attempt by the grains to slide are relaxed by a physical rearrangement of matter. It is the rate at which the latter processes proceed that is often thought to govern the rate at which strain is accumulated.

4.3.1 Dislocation Models

When grain boundaries slide, stress concentrations develop wherever that sliding is obstructed. Relaxation of the stress concentrations by the emission of dislocations from one grain boundary and their absorption by another can be limited by the rate at which the dislocations are emitted (source control), the rate at which they can cross the grains (glide or lattice climb control) or the rate at which they are absorbed into the boundaries (grain boundary climb control). The glide process has been assumed to occur relatively rapidly in superplasticity since there is a lack of either 'strong' obstacles or significant solute drag effects within the grains at the deformation temperature [38,127]. Pile-ups of dislocations adjacent to the grain boundaries are thought to develop and provide a back stress against which the sliding grain(s) would have to work to emit further dislocations along a particular slip plane. Climb of the leading dislocation from the pile-up into the boundary would allow another dislocation to be emitted and enable a small increment of grain boundary sliding (≡ strain) to be accumulated (Fig. 4.3). Stress exponents and grain size dependencies of 2 would be predicted together with an activation energy commensurate with that of grain boundary diffusion. If only one source per grain were activated then flow would be boundary climb controlled. By allowing grain boundary ledges to act as dislocation sources, the absolute magnitude of the predicted strain rate can vary by upwards of two orders of magnitude and is therefore source dependent. The deformation rate is given by

$$\dot{\varepsilon}_{xx} = A_{xx} \frac{D_{gb}\delta Gb}{kT} \left[\frac{b}{d} \right]^2 \left[\frac{\sigma}{G} \right]^2 \qquad (4.7)$$

An alternative view of dislocation accommodated flow arises if grain boundary sliding is accommodated by dislocation climb and glide within the grain boundaries themselves [143,144]. Pile-ups of grain boundary dislocations could form in the grain boundaries at triple points.

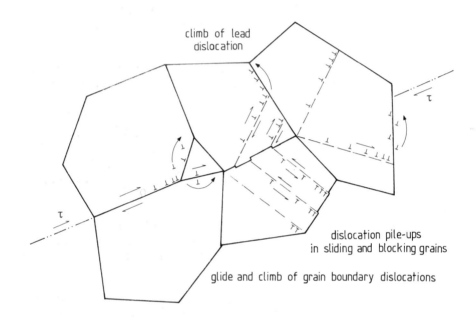

Figure 4.3 Dislocation pile-up models of
superplastic flow. The rate at which the grains
slide past each other can be controlled by (i) the
removal of a pile-up of lattice dislocations (Ball
and Hutchison [38]), (ii) The emission of lattice
dislocations from grain boundary ledges (Mukherjee
[127]) and (iii) by the removal of pile-ups of grain
boundary dislocations (Gifkins [131], Falk et al.
[144]).

Dissociation of the leading dislocation into either lattice dislocations, or boundary dislocations which could glide in the other boundaries intersecting the triple point, would then be the rate controlling process (Fig. 4.3).

Several arguments have been raised against dislocation based models of superplastic flow, namely:

1. The dislocation pile-up models do not predict a threshold stress for superplastic deformation, i.e. Region I.
2. There is no implicit mechanism by which the crystal lattice of either the sliding or accommodating grains could rotate in the lattice pile-up model of Ball and Hutchison [38].
3. Grain elongation is implicit in any model involving dislocation glide/climb on a limited number of slip systems.
4. Dislocation pile-ups are not observed experimentally. Furthermore, at the high temperatures at which deformation takes place pile-ups would not be expected since the average stress is low.

If grain rotation was accepted as resulting from a non-balanced system of unrelaxed grain boundary shear stresses (see Section 4.3.3) then grain elongation would not be observed. Random variations in the direction and magnitude of rotation would cause slip to switch from one slip system to another. Since the rotations can be large (>30°) no net change in grain shape would then be apparent.

Grain rotation would, on the other hand, be an implicit feature of the grain boundary dislocation model. Again, the shear stresses acting on the grain boundaries would be able to move dislocations causing a shear within the boundary zone. The resultant torque on the grains would reorientate the crystal lattice of the grain and an oscillatory grain motion would be observed. Figure 4.4 shows the effect of superplastic strain on the microstructure of a Pb-62Sn eutectic alloy. It is clearly apparent from the micrographs that the grains not only slide

b

a

2 μm

76

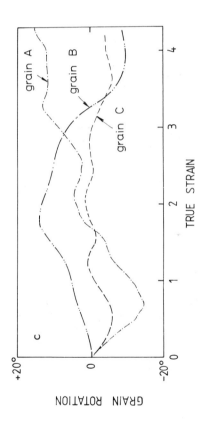

Figure 4.4 The variation of grain orientation with increasing superplastic strain in the Pb-62Sn eutectic. (a) and (b) observed progression of grain sliding with increasing superplastic strain [145] and (c) the measured variation of grain orientation with strain [146].

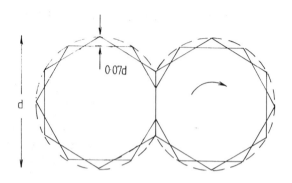

Figure 4.5 Core and mantle model of superplasticity
(Gifkins [131])

and rotate, but also move perpendicular to the free
surface [145]. Measurements of the orientation of
specific grains has shown that the rotations can
vary by as much as 40°, Figure 4.4c [146].

The objection raised against dislocation
accommodated flow based on the absence of either
dislocation activity or dislocation pile-ups can be
countered by the following arguments. Firstly, the
stress generated at the head of the pile-up could
not be supported by the grain boundary at the
deformation temperature and climb would be rapid.
Secondly, on removal of the applied stress there
would be virtually nothing within the
microstructure at that temperature to hold the
dislocations in the piled-up configuration and the
dislocations would run back to their sources.

The grain boundary dislocation model of
superplastic flow is often referred to as the 'core
and mantle' model, as the accommodation of grain
boundary sliding is assumed to occur only within a
viscous mantle around a rigid grain core. If the
grains can be thought of as regular hexagons then
the predicted width of the mantle is only 0.07
times the grain diameter (Fig. 4.5). In a typical

superplastic material the mantle would be only 30
to 70 nm wide!

4.3.2 Diffusion Models

It has also been envisaged that mass could be
redistributed by diffusional flow [147,148]. Driven
by differences in the stress dependent chemical
potential on adjacent grain boundaries, mass
transport from regions of compression to tension
would occur. Sliding is accommodated by a gradual
change in grain shape as matter is moved by
diffusion. Grain boundary migration restores the
original equiaxed shape but in a 'rotated'
orientation (Fig. 4.6). The retention of an
equiaxed grain shape is therefore achieved in the
model of Ashby and Verrall. Furthermore, because a
transient but finite increase in grain boundary
area results from the shape change, the model
predicts a threshold stress for superplastic flow.
The strain rate in the superplastic region is that
due to diffusion but operating under an apparent
stress, σ^*. This stress is equal to the applied
stress less the stress necessary to create that
additional grain boundary energy.

$$\dot{\epsilon}_{I-II} = A_o \frac{D_{diff} G b}{kT} \left[\frac{b}{d} \right]^2 \left[\frac{\sigma^*}{G} \right] \qquad (4.8)$$

If the sliding results in grain switching then no
matter how small the individual three-dimensional
shape change steps are, a finite increase in the
grain boundary area would still be required. The
work done in creating the new grain boundary
surface would be some fraction of the
instantaneous grain boundary area and would have to
be supplied by the external stress. The increased
grain boundary energy is used to drive the
subsequent grain boundary migration but the energy
is lost in the form of heat. The work done per
grain would vary as d^2. The number of grains in
unit volume varies as $1/d^3$ and thus the threshold
stress, which is the work done per unit volume,
will vary as $1/d$. The threshold stress for
superplastic flow predicted by the diffusion

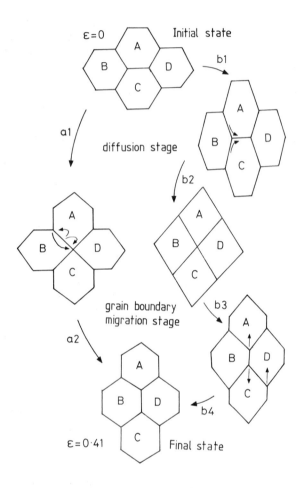

Figure 4.6 Grain switching through diffusional mass
transport as postulated by Ashby and Verrall [147]
(a₁-a₂) and modified by Spingarn and Nix [148]
(b₁-b₂-b₃).

accommodation model varies as the inverse of the grain size.

As with the case of the dislocation based models, several objections have been raised to the accommodation of superplastic flow solely by diffusion, namely:

1. The diffusion paths originally proposed by Ashby and Verrall [147] required that diffusion takes place in different directions on opposite sides of the same grain boundary. As diffusion is driven by the stress acting perpendicular to the grain boundary this is physically impossible (Fig. 4.6).
2. Deformation in the Ashby-Verrall model is not symmetrical.
3. If grain boundary sliding is accommodated solely by diffusion then the lattices of the individual grains cannot rotate. The rotations shown in figure 4.6 are only apparent and result from grain boundary migration.
4. The strain rates predicted by equation (4.8) are about two orders of magnitude too fast.
5. Elongated grains should be apparent in the microstructure.
6. The threshold stress decreases with increasing grain size, contrary to the experimental evidence. Moreover, the predicted threshold stress is significantly less than that measured [263].
7. The grain switching event can only be invoked once giving a maximum strain of 0.55 ($\epsilon_{max}=0.41$ for the two-dimensional model illustrated in figure 4.6).

The diffusion paths were later modified by Spingarn and Nix [148] so that each grain within the cluster underwent the same change, maintaining symmetry of deformation and a more realistic shape transient (Fig. 4.6).

The grain size dependence of strain rate, p, given by eqn (4.8) falls between 2 and 3. However, the observation of a stress exponent of 2 arises only as a result of the transition from the threshold stress (Region I, n=∞), through normal diffusional flow (Region II, n=1) operating under an apparent

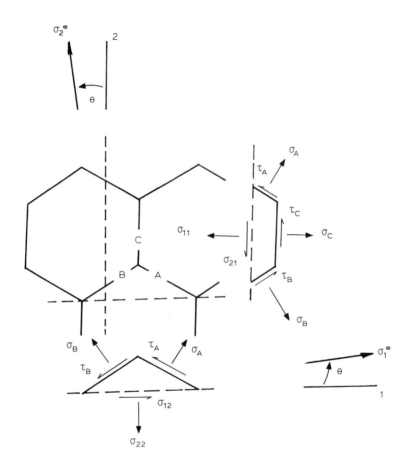

Figure 4.7 An idealised array of hexagonal grains subjected to a uniaxial stress. The stresses on each of the grain boundaries have been resolved into those parallel to the boundary (τ) and those normal to the boundary (σ).

stress, to conventional dislocation creep (Region III, n=5), rather than a distinct superplastic flow process.In view of the independence of both diffusion and dislocation creep processes, it would be more logical if grain boundary sliding was accommodated by both forms of mass transport rather than either in isolation. However, before examining the multimechanism models of superplastic flow, the presumption that grain boundary sliding itself is not a rate controlling process is considered next.

4.3.3 Grain Boundary Sliding

To examine the process of grain boundary sliding consider a two-dimensional array of regular hexagons orientated at random with respect to a remotely applied stress field (Fig. 4.7). The normal and shear stresses acting on each grain boundary can be balanced by stresses acting throughout the continuum body. The magnitude of the shear stresses or normal stresses can only be evaluated if an assumption is made with regard to one of the types of stress [149]. If it is assumed that the grain boundary shear stresses are fully relaxed ($\tau_a=\tau_b=\tau_c=0$) then the grain boundary sliding rate will be controlled by the rate at which the normal stresses on the grain boundaries are relaxed. For the case of uniaxial tension, the average stress, $\hat{\sigma}$, on a given grain boundary is

$$\hat{\sigma} = \sigma\infty \ (3/2 - 2\sin\theta) \tag{4.9}$$

where $\sigma\infty$ is the applied stress, and θ is the angle between the applied stress and the normal to the grain boundary. If, for example, diffusion relaxes the normal stresses then the strain rate is simply that due to diffusion creep, i.e.

$$\dot{\epsilon}_{xx} = 15 \ \frac{D_{diff}Gb}{kT} \ \left[\frac{b}{d} \right]^2 \left[\frac{\sigma}{G} \right] \tag{4.10}$$

This may be compared with eqn (4.8) except that there is no threshold stress. During diffusional stress relaxation, the local value of the normal stress can vary from 0 at the triple points to anywhere between +2.16 and -0.72 times the applied stress at the centre of the grain boundaries. For uniaxial compression, diffusional mass transport would be predicted to occur from the centre of grain boundaries orientated approximately perpendicular to the applied stress, via the triple point, to the centre of boundaries orientated mainly parallel to the applied stress. If the applied stress was tensile then diffusional flow would occur in the opposite direction. The predicted diffusion paths are in excellent agreement with the modified paths for diffusion accommodated superplastic flow [148] (Fig. 4.6).

If on the other hand the normal tractions on the grain boundaries are relaxed by rapid diffusional flow ($\sigma_a = \sigma_b = \sigma_c = \sigma/2$) then the average shear stress, $\hat{\tau}$, acting parallel to the boundary would be given by

$$\hat{\tau} = -\sigma\infty \,(2\sin\theta.\cos\theta) \tag{4.11}$$

and the corresponding strain rate by

$$\dot{\epsilon}_{xx} = A \left[\frac{\sqrt{3}}{2} \right]^{n-1} \left[\frac{\sigma}{G} \right]^n \tag{4.12}$$

where A is a rate constant, and n the stress exponent characterising the stress relaxation process. If all the boundaries are equally resistant to sliding then the strain rate is independent of the orientation of the grains. If the boundary viscosity varies with, for example, misorientation or precipitate density, then

$$\dot{\epsilon}_{xx} = \frac{1}{\sqrt{3}d} \left[(\dot{S}_a - \dot{S}_b)(2\cos^2\theta - 1) \right.$$

$$\left. + (\dot{S}_a + \dot{S}_b - 2\dot{S}_c)(2\sin\theta.\cos\theta) \right] \tag{4.13}$$

where the grain centre shear rate, $\dot{S}_\perp$, is

equal grain centre offset

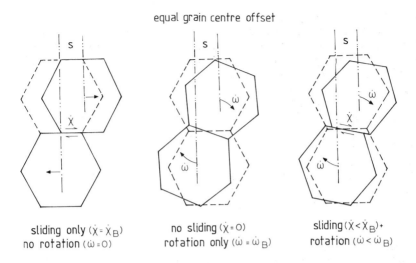

sliding only $(\dot{\chi}=\dot{\chi}_B)$
no rotation $(\dot{\omega}=0)$

no sliding $(\dot{\chi}=0)$
rotation only $(\dot{\omega}=\dot{\omega}_B)$

sliding $(\dot{\chi}<\dot{\chi}_B)$+
rotation $(\dot{\omega}<\dot{\omega}_B)$

Figure 4.8 Sliding and rotational displacements required to generate a given grain centre offset between two adjacent hexagonal grains. (a) sliding only (X_B), (b) rotation only (w_B) and (c) a mixture of sliding and rotation ($X<X_B$ and $w<w_B$).

related to the grain boundary shear rate, $\dot{X}_i$, and the shear stress acting on that boundary, τ_i, through the relationships

$$\dot{S}_i = \dot{X}_i \quad \text{and} \quad \dot{X}_i \propto \tau_i^n \qquad (4.14)$$

The grain boundary shear rate is only equal to the grain centre shear rate when the grains do not rotate. Grain boundary sliding can, therefore, be accommodated by relaxing the shear stresses in and adjacent to the grain boundaries, and the process can be equated with the 'core and mantle' model of superplastic deformation [131].

From figure 4.8, it can be seen that a given grain centre shear displacement, S, and hence a

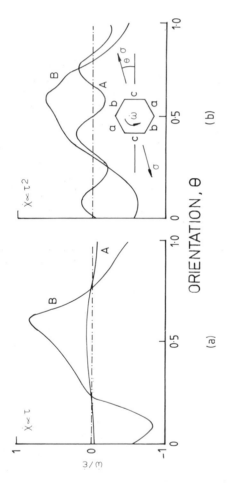

Figure 4.9 The variation of grain rotation rate ($\dot{w}$) with grain orientation (θ) when the grain boundary sliding rate ($\dot{X}$) varies as (a) τ and (b) τ^2, when the grain boundary viscosities differ by 10% (curve A) and by 90% (curve B) [149].

86

macroscopic strain, ϵ, can be achieved not only by sliding a distance X_B along the boundary, but also by rotation of the boundary through an angle w_B. If both processes are allowed to operate simultaneously then a given grain centre shear displacement (rate) can be achieved by lesser rotation and sliding (rates), i.e.

$$\dot{S} = \dot{X} + \dot{w}d \qquad (4.15)$$

where $\dot{w}$ is the average rotational velocity of the grains on either side of the grain boundary. If no holes appear in the structure then the average strain rate is twice that when rotation is not allowed [149]. A given strain rate can therefore be maintained for smaller grain boundary sliding rates and hence at a lower stress. Moreover, for a given grain boundary sliding rate the strain rate would increase as the grain size was reduced. Thus grain boundary sliding and superplastic deformation would be expected to become increasingly apparent in progressively finer grain materials.

The optimum combination of grain boundary sliding and rotation can be evaluated by minimising the rate of doing work, W, where

$$W = \sum_i \dot{X}_i \tau_i$$

and

$$\frac{dW}{dX} = 0 \qquad (4.16)$$

The variation of the angular rotation rate, normalised against the strain rate has been calculated by Beeré [149] and is shown in figure 4.9 as a function of the grain boundary orientation and relative grain boundary viscosities. The momentary rate of grain rotation can be quite high, though the maximum rotation is seldom greater than 30° (see also figure 4.4). Thus it seems that the kinetics of grain boundary sliding are controlled by the slower of two accommodation processes, and grain boundary sliding would appear not to be a unique mechanism of superplastic deformation but rather a natural consequence of an unbalanced rate of grain boundary shear stress relaxation.

From the preceding section it is clear that the normal and shear stresses generated on the boundaries in response to an attempt by the grains to slide, can be relaxed by both diffusion and dislocation glide/climb. The stresses on the grain boundary will vary both with orientation and with grain boundary structure. Variations in the kinetics of stress relaxation around a grain or group of grains can explain the topological features of superplastic flow. The macroscopic flow characteristics will therefore be governed not only by the absolute and relative magnitudes of the individual stress relaxation processes, but also by the way in which they combine.

Evidence for what has come to be known as 'micromultiplicity', has been provided by in situ studies of superplastic deformation where the deformation process itself has been observed to be highly heterogeneous. Several deformation mechanisms would appear to operate simultaneously or sequentially within different regions of the microstructure. Indeed such an explanation of superplasticity provides a relatively simple rationale for the wide divergences in the reported values of strain rate sensitivity, activation energies and Region I behaviour. However, developing such a concept into a fully quantitative theory of superplastic flow would be exceedingly complex.

4.3.4 Transitional Models

The observation of similar 'm' values and activation energies in Regions I and III has led several researchers to regard the superplastic region as a transitional one rather than a deformation process in its own right. So far, both the qualitative and quantitative interpretation of the observed characteristics of superplasticity has been considered mainly in terms of a single phase material defined by a unique value of a microstructural parameter, namely the grain size. However, only two types of single phase materials have actually been the subject of sustained experimental investigation and the majority of the theories of superplastic flow have been derived from measurements obtained from duplex materials. The single phase alloys for which a wide range of experimental measurements are available are two aluminium-base alloys Al-7475 and Al-2004 (Supral 220) and a copper-base alloy CDA-638. The experimental data for the copper alloy is particularly interesting and covers a wide range of strain rates, temperatures, grain sizes and testing techniques [109,110,150]. It is the effect of grain size on the shape of the stress-strain rate relationship that would appear to offer the greatest clue to exposing the true identity of Region II.

No evidence of a threshold stress for deformation in Region I has been obtained in either the copper or the aluminium base alloys. The region of low strain rate sensitivity would appear to extend to strain rates less than 10^{-7}/s at a constant slope (Fig. 4,10). Moreover, the activation energy for flow in Region I, measured from both isothermal and temperature cycling tests, was the same as that in Region III; both were about 10% lower than that for volume diffusion in pure copper. The activation energy for flow in Region II was lower than that in Regions I and III but about 15% greater than that for grain boundary diffusion in copper. Increasing the grain size from 2.7 μm to 100 μm resulted in a decrease in the slope of Region II, the two regimes of low strain rate sensitivity ultimately becoming continuous (Fig. 4.10).

The effects of temperature and grain size on the stress-strain rate characteristics of several commercial aluminium alloys have also been investigated. In general, decreasing the temperature and increasing the strain rate had the same effects as in the copper alloy CDA-638, i.e. Region III extended to lower strain rates at the expense of Region II. However, very little data exists for strain rates less than 10^{-5}/s where

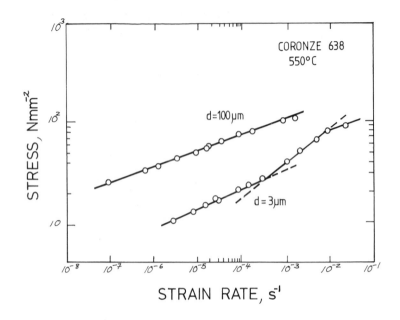

Figure 4.10 Plot of stress versus strain rate for CDA 638 at 550°C showing a distinct superplastic region when d=3μm but an unpunctuated transition from Region I to Region III when d = 100μm [111].

the alloys investigated still appeared to have a high strain rate sensitivity. Unfortunately, the aluminium alloys examined were microstructurally unstable and strain enhanced grain growth would make the acquisition of meaningful data at lower strain rates very difficult.

The experimental studies on CDA-638 show that the stress exponents for Regions I and III are the same and lie between 3 and 5. It follows that the low stress process should also be recovery controlled dislocation creep. In the fine grain material, however, it differs from that in Region III by the absence of a dislocation structure within the

90

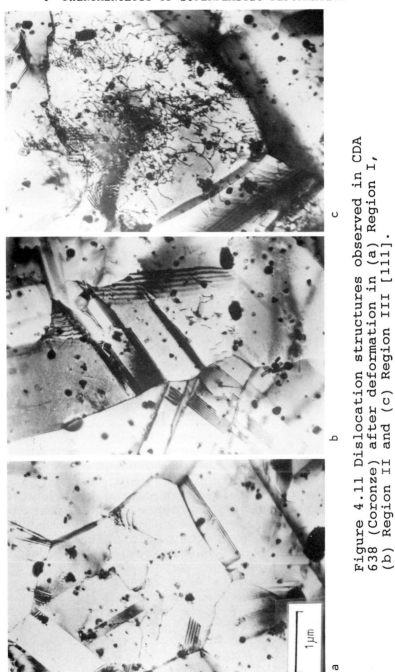

Figure 4.11 Dislocation structures observed in CDA 638 (Coronze) after deformation in (a) Region I, (b) Region II and (c) Region III [111].

grains (Fig. 4.11). The equilibrium subgrain size, d_s, can be calculated from the relationship

$$d_s = 20b \left[\frac{\sigma}{G} \right]^{-1} \qquad (4.17)$$

At the strain rates associated with Region I deformation, the subgrain size is greater than the grain size of many superplastic alloys. Dislocation glide across and around the grains is impeded only by the relatively weak forest dislocations, solutes and any precipitates remaining within the grains at the deformation temperature. The deformation process is either controlled by climb within the grains (lattice diffusion) or by climb in and adjacent to the boundaries (boundary diffusion). Assuming that the latter process is the more important then it has been shown that the strain rate in Region I, where d>d_s is given by [151]

$$\dot{\varepsilon}_I = A_I \frac{D_{gb}\delta G}{kT} \left[\frac{b}{d} \right]^1 \left[\frac{\sigma}{G} \right]^5 \qquad (4.18)$$

However, increasing the strain rate or stress results in the magnitudes of the grain size and subgrain size becoming compatible. Dislocation tangles encroach from the grain boundaries into the grains, and subgrain boundaries begin to become apparent. As the stress is increased further, so the dislocation networks progress toward the centre of the grains and plastic flow gradually tends toward conventional recovery controlled creep. The superplastic regime is therefore a transition from a region where d=d_s to one where d>>d_s. If the strain rate in Region III is given by:

$$\dot{\varepsilon}_{III} = A_{III} \frac{D_v Gb}{kT} \left[\frac{\sigma}{G} \right]^5 \qquad (4.19)$$

then the strain rate in the transitional regime,

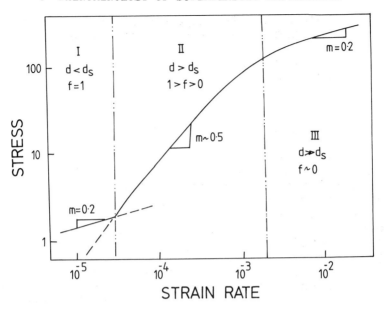

Figure 4.12 Schematic illustration of the $\sigma:\dot{\epsilon}$ curve predicted by the sub-grain transition model of Spingarn and Nix [148,151]. (d_s, sub-grain diameter; d, grain diameter)

Region II, where $d>d_s$ is

$$\dot{\epsilon}_{II} = f\dot{\epsilon}_I + (1-f)\dot{\epsilon}_{III} \qquad (4.20)$$

where f is the ratio of the subgrain size to the grain size. The variation of stress with strain rate predicted by the transitional model is shown in figure 4.12, from which it can be seen that the strain rate sensitivity of the transitional regime is greater than in either Regions I or III. It also follows from eqns (4.18) to (4.20) that the values of m,p and Q_{SP} will vary throughout Region II. Unfortunately, the slip transition model predicts an activation energy for superplastic flow that is equal to that for grain boundary rather than lattice diffusion. Similarly, a grain size exponent of 0 to 1 would be expected in Region II rather than a value of 2 to 3 which is observed experimentally.

An alternative explanation for the apparent similarities between deformation in Regions I and III relies on the observation that the quoted grain size of a material is in reality the mean value of a distribution of grain sizes, or that the physical properties of the material are the mean of those of two distinctly different phases. The variation of stress with strain rate would depend on the relative proportions of the two phases, or the form of the distribution of grain sizes and on whether the response of the 'composite' was that obtained by an isostress or isostrain rate rule of mixtures, i.e.

$$\sigma = \sum_i f_i \sigma_i (\dot{\epsilon}) \qquad (4.21)$$

for an isostrain rate model, or

$$\dot{\epsilon} = \sum_i f_i \dot{\epsilon}_i (\sigma) \qquad (4.22)$$

for an isostress model, (Fig. 4.13) [152].

The transitional model for superplastic flow based on the distributed microstructure and isostrain rate rule of mixtures is attractive from two standpoints. Firstly, only the two conventional high temperature deformation processes are required and, depending on the relative proportions of the two microstructural extremes (fine and coarse grains) or the divergence of properties of each of the two phases, any value for the stress exponent (n=1/m) from 1 to 5 can be generated for the transitional (superplastic) region and maintained over two to three orders of magnitude of strain rate. Secondly, since the transition region is dominated by diffusive stress relaxation, the flow

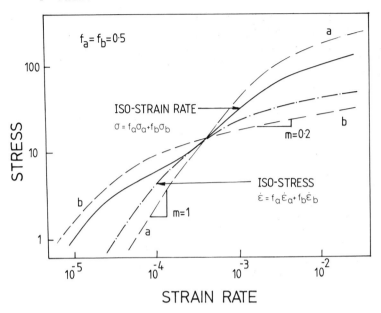

Figure 4.13 Schematic illustration of the $\sigma:\dot\epsilon$ curves predicted by the rule of mixtures model (═══ isostrain rate, -.-.-.- isostress) with only conventional diffusion creep (m = 1) and dislocation creep (m=0.2) operating in each of the two phases of the microstructure (curves a and b show the variation of stress with strain rate for phase a and b, respectively).

stress would be highly grain size dependent. However, neither transitional hypothesis provides an explanation for the topological features of superplastic flow, such as the maintenance of an equiaxed grain shape, that are so characteristic of superplasticity.

4.3.5 Superplastic deformation in two phase alloys

There are large numbers of duplex materials which show a higher degree of superplasticity than the pseudo single phase alloys. The mechanical and topological features of deformation are similar to those of the single phase alloys except that deformation generally takes place at a lower homologous temperature, for example ~0.6T_m for α/β titanium and α/β copper alloys as compared with ~0.85T_m for aluminium alloys. The phenomenological equations describing superplastic deformation which have been given in the previous sections have assumed a single phase microstructure, yet such equations were often derived from measurements made using duplex materials. The material parameters represented in these equations were either those of the softer phase or those given by the rule of mixtures. A much more elegant realisation of the 'average' properties approach is embodied in the principles of the transitional model set out in section 4.3.4, provided an isostrain rate rule of mixtures is used (Fig. 4.13). However, such an assumption would appear contrary to the observation that superplastic flow seems to be accommodated almost exclusively within the softer of the two phases (an isostress combination).

Differences in the material properties of each phase of a duplex microstructure would require that the two phases deform at quite different strain rates if an isostress rule were applied. The material could be considered in terms of a non-Newtonian fluid (i.e. a soft β phase with a stress exponent greater than unity) containing hard undeformable particles (for example the α phase in both α/β titanium and α/β copper alloys). Grain boundary sliding would then occur by shear within a thin layer of the β phase in the vicinity of the grain boundaries. The gliding dislocations in the β phase would be more effectively obstructed by the α/β interphase boundaries than by the β/β boundaries. The dislocations might therefore be expected to pile up in front of the α grains, climb along the interphase boundary (IPB) until glide within the β phase could continue. The probability of a gliding dislocation intersecting an α grain would increase as the volume fraction of the harder α phase increased, while the glide distance would depend on the β phase size. The strain rate would then be given by [152]

$$\dot{\epsilon}_{II} = A_{II} \left[\frac{1 - f_\alpha}{f_\alpha} \right]^2 \frac{D_{IPB}G}{kT} \left[\frac{b}{d_\beta} \right]^2 \left[\frac{\sigma}{G} \right]^2 \qquad (4.23)$$

where f_α is the volume fraction of α phase and d_β, the grain size of the β phase. Equation (4.23) is not dissimilar to that predicted by the earlier analysis of Gittus [153] where dislocation climb within the interphase boundary was assumed to be the rate controlling process. Again a stress exponent of 2 is predicted since the model invokes climb from a dislocation pile-up as the rate limiting step in the deformation process.

Support for a model of superplastic flow in duplex materials based on accommodation solely within the softer of the two phases has been provided by the analysis of Chen [154]. He showed that a better agreement between the experimentally determined flow behaviour of Ti-6Al-4V and that predicted from theory could be obtained if diffusion in the two phase alloy was governed by interdiffusion ($D = C_bD_a + C_aD_b$) rather than ambipolar diffusion ($D = D_aD_b/(C_bD_a + C_aD_b)$), i.e. the fastest rather than the slowest diffusing species governs the overall rate of accommodation.

4.3.6 Continuum Model

Chen has also proposed a continuum mechanics model of superplastic flow based on the transformation plasticity theories of Eshelby [155,156]. It was assumed that a grain switching event similar to that envisaged by Ashby and Verrall and others [147-149] was the strain producing step, but that these events occurred stochastically rather than simultaneously at all points within the microstructure. The grain switching events were not only assumed to occur at random in space and time,

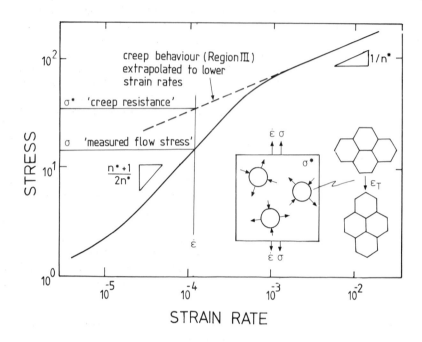

Figure 4.14 Schematic illustration of the $\sigma:\dot{\varepsilon}$ curve predicted by the transformation plasticity model of Chen [156]. The 'local' strain generating event is that postulated by Ashby and Verrall [147] which creates randomly orientated regions of stress which are relaxed by macroscopic flow in the direction of the applied stress.

but also with respect to the direction of plastic
flow. Since the individual grain switching events
were totally random, after a sufficiently large
number of events there would be no net strain,
since they would cancel the effect of each other
out. However, following each grain switching event,
the reorientated grains would still have to fit
within their original locations. To accommodate the
switched grains in their new configuration, a large
localised concentration of strain develops in the
surrounding matrix. The local strain would be
relaxed by plastic flow throughout the body of the
material and the direction of that relaxation would
be biased by the applied stress in the macroscopic
flow direction.

The local strain, generated by the grain switching
event, then becomes redistributed and vanishes but
causes a very small displacement of the outer
surfaces of the material in the flow direction. The
relaxation of the local stresses produced by
subsequent grain switching events, regardless of
their orientation, each contribute a little to flow
in the direction of the applied stress (Fig. 4.14).
The resultant constitutive equation for
superplastic flow, assuming that the long range
stress relaxation process is recovery controlled
dislocation creep, is given by

$$\dot{\epsilon}_{xx} = A \; \epsilon_{tr}^{n/(n+1)} \; \frac{D_{diff}Gb}{kT} \left[\frac{b \; \sigma}{d \; G} \right]^{2n/(n+1)} \tag{4.24}$$

where ϵ_{tr} is the unconstrained transformation
strain associated with the grain switching event,
and n is the stress exponent of the long range
relaxation process (in this case n = 5).

The continuum interpretation does not consider the
detailed mechanism of the grain switching process
since that transformation does not in itself
control the rate at which a macroscopic strain is
accumulated. As the macroscopic strain is
accumulated only from the long range relaxation of
the strains (or stresses) generated by that
switching event then the orientation of that event

is irrelevant. Grain switching can occur repeatedly back and forth at the same point within the microstructure and each time generates a small strain in the macroscopic flow direction. The model does not, therefore, suffer from the same restrictions as that of Ashby and Verrall and of Spingarn and Nix [147,148]. It can be seen from eqn (4.24) that the stress and grain size exponents are non-integer yet close to 2, and that the activation energy is somewhat greater than that of grain boundary diffusion.

4.4 Summary

In the preceding sections it has become apparent that no single process can explain completely the mechanical aspects of superplastic flow, as manifested in the measured stress-strain rate characteristics. Nor could it account for the topological aspects of superplastic flow, namely the maintenance of an equiaxed grain shape, grain rotation and translation. Intuitively this would be expected since numerous processes are simultaneously operative within a microstructure which is far from uniform. Nonetheless, it has been demonstrated that grain boundary sliding becomes increasingly significant in finer grain materials. Superplastic flow may be divided into two parts, microscopic and macroscopic.

Microscopic aspect

In the absence of any significant intragranular dislocation structure, grain boundary sliding and grain rotation are accommodated by diffusive mass transfer and dislocation glide/climb over extremely short path lengths. The direction and rate of the various accommodation processes will be sensitive to the local geometry, grain boundary structure and chemistry, and could not be predicted with any degree of certainty, especially as the way in which the various mechanisms would combine is not known. However, it is such processes that are responsible for the topological aspects of superplasticity.

Macroscopic aspect

In view of the difficulties involved in the development of a fully quantitative theory of superplasticity based on the micromultiplicity approach it may be more profitable to look at the behaviour of the continuum body in response to such 'grain switching' events. The transformation plasticity hypothesis does not require a detailed understanding of the grain switching process except that it does occur and results in a localised concentration of strain. The kinetics of the long range relaxation of that strain throughout the continuum, biased by the applied stress, define the rate of displacement of the external surfaces of the material, and hence the perceived macroscopic stress-strain rate behaviour at a given temperature for a given mean grain size.

5
CAVITATION AND FRACTURE

5.1 Introduction

When a superplastic material fails during tensile deformation it is either the result of unstable plastic flow or a consequence of the growth and interlinkage of internally nucleated voids. In the former process, inhomogeneities in the cross-sectional area of a test piece lead to a localised increase in the strain rate and the difference in the cross-sectional area increases. The rate at which the discontinuity in the cross-sectional area increases depends on both the rate of strain hardening and the strain rate sensitivity [157-159]. In superplastic materials, where true strain hardening is minimal, any neck which is present will always grow, although the rate of growth decreases with increasing m. Unstable plastic flow normally results in the material pulling out to a fine point prior to failure (Fig. 5.1). Where failure occurs as the result of the nucleation, growth and interlinkage of internal voids, the fracture surface is much more abrupt. The value of the strain rate sensitivity is important in determining the rate at which the voids grow and thus to some extent controls the elongation to failure in systems which undergo cavitation.

Figure 5.1 Shadowgraphs of the fracture in two
superplastic alloys after 900% tensile elongation.
(a) unstable plastic flow in Ti-6Al-4V and (b)
pseudo-brittle fracture by cavitation in Supral
220.

5.2 Fracture by Unstable Plastic Flow

The value of the strain rate sensitivity index,
'm', has been shown to have a strong effect on the
ductility of superplastic materials (Fig. 5.2). In
general, the higher the 'm' value, the greater the
elongation to failure [160], although this rule is
by no means universal. Rai and Grant [46], for
example, observed that the strain rate at which the
maximum elongation occurred in a superplastic Al-Cu
eutectic alloy did not coincide with that at which
the maximum strain rate sensitivity was measured
(optimum strain rate). It has been suggested [161]
that the elongation at failure depends on the 'm'
value at high rather than low strains, and that
microstructural evolution during deformation can
have a marked effect on the strain rate
sensitivity.

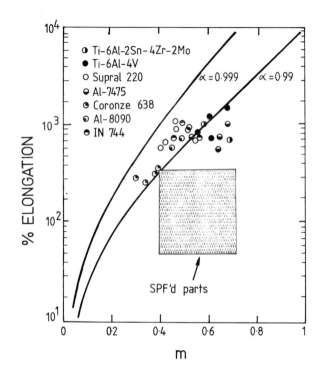

Figure 5.2 The predicted uniform strain to failure as a function of strain rate sensitivity together with corresponding data for several commercial superplastic alloys. The shaded area marks the range of strains that would be required in a superplastic forming process and the m-values are those observed in commercial materials.

Several relationships defining the strain at failure during superplastic flow have been derived [161-167]. The total tensile elongation to failure of a superplastic material will be comprised of both the uniform strain and that accumulated within any necks which develop during deformation.

Consider now a tensile specimen in which there is a small neck for which the initial cross-sectional area varies from a_o to αa_o, where $\alpha < 1$. If the material obeys the flow law then

$$\dot{\epsilon} = k' \sigma^{1/m} \tag{5.1}$$

The load on the tensile test piece is then

$$p = \sigma a \tag{5.2a}$$

and the strain rate

$$\dot{\epsilon} = -(da/dt)/a \tag{5.2b}$$

Hence

$$k' p^{1/m} = - a^{(1/m-1)} \, da/dt \tag{5.2c}$$

It can be seen from equation (5.2c) that as m increases, the rate of change of the cross-sectional area becomes less dependent on the magnitude of that area and any neck would grow more slowly. If, after a small increment of strain, the cross-sectional area of the test piece is a, and that of the neck is βa, where $\beta < 1$, then rearranging equation (5.2c) and integrating gives [162]

$$\int_t^{t+\delta t} k' p^{1/m} \, dt = - \int_{a_o}^a a^{(1/m-1)} \, da \tag{5.3a}$$

and

$$\int_t^{t+\delta t} k' p^{1/m} \, dt = - \int_{\alpha a_o}^{\beta a} a^{(1/m-1)} \, da \tag{5.3b}$$

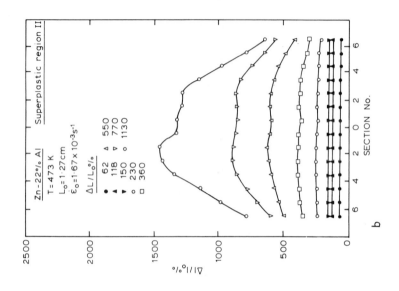

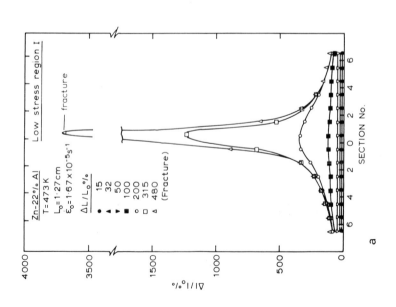

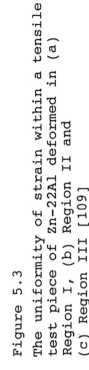

Figure 5.3

The uniformity of strain within a tensile test piece of Zn-22Al deformed in (a) Region I, (b) Region II and (c) Region III [109]

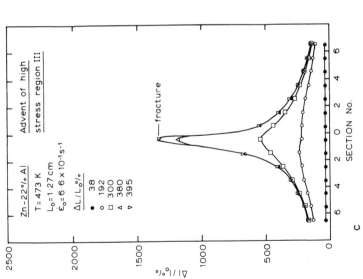

Eliminating the LHS of equations (5.3a) and (5.3b) we obtain

$$a = \left[\frac{1 - \alpha^{1/m}}{1 - \beta^{1/m}} \right]^m a_o \qquad (5.4)$$

The uniform strain in the test piece is simply $\epsilon_u = -\ln(a/a_o)$ and the maximum elongation to failure can be obtained by setting β equal to zero, i.e.

$$\epsilon_u = - m \ln(1-\alpha^{1/m}) \qquad (5.5)$$

If the initial discontinuity is small then there is a very strong dependence of the elongation to failure on the strain rate sensitivity, m. The variation of the elongation to failure vs strain rate sensitivity is plotted in figure 5.2 for two values of α: 0.99 and 0.999. It can be seen that the elongations attained for several commercial superplastic alloys lie towards the lower bound predicted by equation (5.5), but these are still in excess of the elongations that would be required of the material for superplastic forming (shaded area).

The uniformity of plastic flow has been measured in Zn-22Al by dividing the gauge length of a tensile specimen into a number of segments and monitoring the elongation of each segment [116]. It can be seen from figure 5.3 that, for deformation in both Regions I and III, once a neck forms it propagates rapidly. However, for deformation in Region II, where true superplastic flow occurs, plastic flow is much more uniform, even to very high strains, and any necks which do form tend to be rather diffuse.

In the analysis of plastic flow developed above, it has been assumed that the material is structurally stable and the value of m remains constant. Experimental work has shown that the strain rate sensitivity can vary significantly during superplastic flow [48]. The actual variation of m with strain will depend on the (complex) interrelationship between strain rate, temperature, grain growth and strain hardening/softening. For example, in Supral 220 grain growth during deformation causes not only a reduction in the magnitude of the maximum value of the strain rate sensitivity with increasing strain but also leads to lower value of the strain rate at which that maximum occurs. If the deformation rate is constant and higher than the strain rate at which m is initially a maximum, then the strain rate sensitivity will continually decrease with increasing strain (Fig. 5.4). However, if the deformation rate is initially less than the strain rate for maximum 'm' then the strain rate sensitivity would increase during the initial stages of superplastic flow, reach a maximum and then decrease. The actual magnitude of the increase in the strain rate sensitivity depends on the interrelationship between microstructural instability at the deformation temperature and strain rate.

The maximum tensile elongation attainable in superplastic materials is of little relevance when compared to the actual strains which the material would be required to achieve during a commercial superplastic forming operation, but nevertheless provides a useful guide to the superplastic deformation potential of the material. Considering only the uniform strain predicted for a given value of m, most superplastic materials would appear to have ample reserves of 'stretchability'. Unfortunately, not all superplastic alloys pull down to a fine point at failure. Two different modes of failure in superplastic tensile test pieces are shown in figure 5.1. Both samples attained the same strain at failure, but the flat fracture surface of the lower specimen, termed pseudo-brittle, arises from the ductile tearing of tiny ligaments between regions of internal cavitation. It is the development of such cavitation damage that often leads to the premature failure of superplastic materials [168] which will be considered in the remainder of this Chapter.

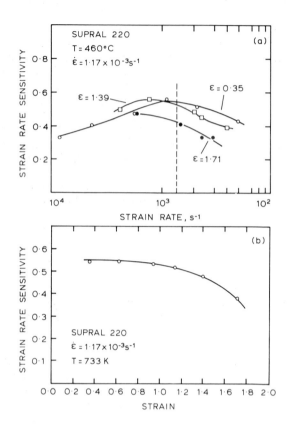

Figure 5.4 (a) The variation of strain rate
sensitivity with strain rate after deformation to
various strains at a strain rate of 1.17×10^{-3}/s
(b) The variation of strain rate sensitivity with
strain during deformation at a strain rate of
1.17×10^{-3}/s (Supral 220).

5.3 General Characteristics of Cavitation

Despite the large plastic strains which can be obtained in superplastic materials it is now well established that cavitation may occur during superplastic flow. The alloys in which cavitation has been observed include those based on aluminium [39,170-174], copper [105,175-181], iron [63,66,70,71,182], lead [183], silver [184], titanium [185-187] and zinc [188-191]. The subject of cavitation in superplasticity has been reviewed extensively both from an experimental and theoretical viewpoint [192-196]. In general, cavities nucleate at the grain boundaries and their subsequent growth and coalescence invariably leads to premature failure. However, and more importantly from a practical viewpoint, the presence of cavities in superplastically formed components could adversely affect their mechanical properties. There is also evidence that cavities may develop from defects which pre-exist, and which are usually produced during the thermomechanical processing required to develop a superplastic microstructure [195].

An important requirement for cavitation during superplastic flow is the presence of a local tensile stress. Under the conditions of homogeneous compression cavitation is not observed, and cavities which are produced during superplastic tensile flow are removed during subsequent compressive flow [197]. Superplastic closed die isothermal forging of Ni-base superalloys such as IN100, starting from hot pressed powders which are heavily worked by extrusion, is used in the manufacture of turbine discs to give a sound cavity free product of uniform microstructure. It has also been demonstrated that the superimposition of hydrostatic pressure during both uniaxial and biaxial superplastic tensile flow can reduce or eliminate cavitation [198]. The cavitation damage which develops during superplastic forming can also be removed by a post-forming hot isostatic pressing (HIPping) treatment which will sinter   the voids [199]. In order to control cavitation it is therefore necessary to understand both the

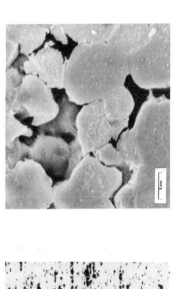

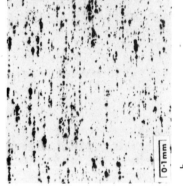

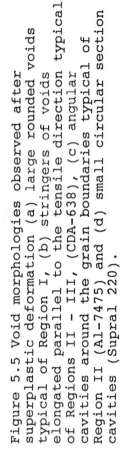

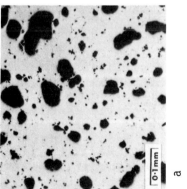

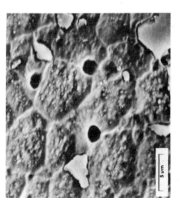

Figure 5.5 Void morphologies observed after
superplastic deformation (a) large rounded voids
typical of Region I, (b) stringers of voids
elongated parallel to the tensile direction typical
of Regions II - III, (CDA-638), (c) angular
cavities around the grain boundaries typical of
Region II (Al-7475) and (d) small circular section
cavities (Supral 220).

112

microstructural and deformation parameters which influence its occurrence and magnitude.

In the majority of studies on cavitation only the variation of the total volume fraction of cavities with strain at different strain rates and temperatures has been studied. It should be noted that the volume fraction of cavities, C_v, can be written as

$$C_v = \Sigma \; N_{vi} . V_i \tag{5.6}$$

where N_{vi} is the number of cavities per unit volume having a volume V_i. The number of voids of any particular size is related to the number of pre-existing voids and to the nucleation rate, while the volume of a particular void is controlled essentially by the growth rate. However, during the latter stages of deformation when the cavitation level is high (approaching 10%), the growth process is affected by the spatial distribution of the voids through void coalescence. Hence growth becomes, in part, dependent on the nucleation process. It is therefore evident that the variation of the volume fraction of voids with strain, strain rate and temperature, will not be a simple one. The effect of changing the strain rate and temperature at which deformation occurs may be further complicated by other factors such as grain growth and/or a change in phase proportions [105,175].

The morphology of cavities formed during superplastic flow varies from one material to the next and even in the same materials deformed at different strain rates. In general, three types of cavities have been observed in superplastic materials (Fig. 5.5). These are:

1. Spherical voids with radii up to ~100 μm.
2. Elliptical voids elongated parallel to the tensile axis with lengths up to ~50 μm and aspect ratios from 2:1 to 10:1.
3. Groups of angular or crack-like cavities, each up to 10 μm in length, interlinked around clusters of grains.

The cavity morphologies have been cited as evidence for the operation of different void growth mechanisms. The circular section voids have often been taken to infer diffusional growth, while the elongated elliptical section voids are thought to be indicative of strain controlled growth. However, there have been few systematic metallographic studies (or quantitative analyses) of the evolution of void size and shape during superplastic flow.

Unfortunately, observations of cavities at the nucleation stage are usually difficult to make. Cavities do not form uniformly throughout the microstructure and information concerning nucleation has to be deduced from specimens in which the voids have grown to a size where they are resolvable, either by direct observation or by the measurement of a physical property of the base material. Since cavities do not always nucleate during superplastic flow and, with only very specific exceptions, will always grow during tensile superplastic deformation, the mechanisms of cavity growth will be examined prior to a consideration of the nucleation process.

5.4 Cavity Growth

5.4.1 Diffusion controlled growth

A cavity located on a grain boundary, whether nucleated during superplastic flow or pre-existing, may grow by stress directed vacancy diffusion, by plastic deformation of the surrounding matrix or by a combination of both these mechanisms. The former process has been analysed by several workers [200-209] for the case of spherical voids under uniaxial tension and leads to a relationship of the form

$$\frac{dr}{d\epsilon} = \frac{D_{gb}\delta\Omega}{kT\dot{\epsilon}r^2} \left[\sigma_1 - \frac{2\Gamma}{r} \right] \frac{1}{\ln(1/2r) - 3/4} \qquad (5.7)$$

where σ_1 is the maximum principal stress local to the grain boundary, Ω is the atomic volume, Γ, the surface energy and r the cavity radius. The

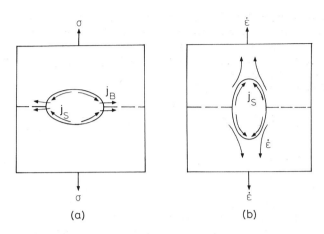

Figure 5.6 Schematic illustration of the cavity (void) growth process for (a) diffusion controlled growth, (b) strain controlled growth.

driving force for cavity growth is provided by the difference in chemical potential between an atom on the stressed grain boundary and that on the free surface of the void. When the maximum principal stress is tensile, atoms move from the void surface to the grain boundary and the void will grow (Fig. 5.6). If the stress is compressive then atoms move in the opposite direction and the voids will sinter.

The relationships derived in the literature differ only in the form of the last term containing solely r and l, the size and spacing of the voids. In deriving relationships of the form of equation (5.7) it has been assumed that the voids are widely spaced on fixed boundaries orientated perpendicular to the applied stress axis and are themselves small relative to the grain size of the material. It is evident from equation (5.7) that the rate of void growth decreases as the inverse square of the cavity radius and thus slows considerably as the voids grow. In the majority of superplastic materials, the voids have dimensions similar to that of the grain size. Furthermore, the cavities are usually located on sliding boundaries which are orientated randomly with respect to the applied stress. It was proposed by Miller and Langdon that when the void diameter was the same as the grain size, then the growth rate would be enhanced by additional mass transfer along boundaries intersecting the surface of the void [210]. Such enhancement was later shown to be sensitive to the ratio of the void size to the grain size [211]. The rate of void growth is then given by [212],

$$\frac{dr}{d\epsilon} = 45 \frac{D_{gb}\delta\Omega}{kT \, d^2} \left[\frac{\sigma_1}{\dot{\epsilon}} \right] \qquad (5.8)$$

where d is the grain size. Unlike the previous equation (5.7), the rate of change of void radius with strain is independent of the void radius, r.

Equations (5.7) and (5.8), which were initially derived for the case of uniaxial tension, can be written in more general terms allowing the effects of multiaxial states of stress and alternative geometries of deformation to be considered. (Commercial superplastic forming usually involves balanced biaxial or plane strain rather than uniaxial deformation.) The maximum principal stress, σ_1, can be redefined in terms of the von Mises equivalent stress, σ_E, (where $\sigma_E = 1/\sqrt{2}((\sigma_1-\sigma_2)^2+(\sigma_2-\sigma_3)^2+(\sigma_3-\sigma_1)^2)^{1/2}$) and the superimposed pressure, P, i.e.

$$\frac{\sigma_1}{\sigma_E} = \left[\frac{k_D}{3} - \frac{P}{\sigma_E} \right] \qquad (5.9)$$

where k_D is a geometric factor that depends on the mode of deformation (uniaxial, balanced biaxial or plane strain) and the extent of grain boundary sliding. The determination of appropriate values for k_D will be discussed in section 5.4.3.

5.4.2 Plasticity or strain controlled growth

For cavity growth by deformation of the surrounding matrix in uniaxial tension, the relationship

$$\frac{dr}{d\epsilon} = \frac{\cap}{3}\left[r - \frac{3\Gamma}{2\sigma_E}\right] \qquad (5.10)$$

has been proposed, where $\cap$, the cavity growth rate parameter, is dependent on both the applied stress state and the geometry of deformation [192,193]. Unlike diffusion controlled cavity growth, the rate of void growth increases linearly with void size and is independent of the strain rate. Plasticity dominated void growth is controlled by the mean stress ($\sigma_M = (\sigma_1+\sigma_2+\sigma_3)/3$) rather than the maximum principal stress.

The parameter $\cap$ in equation (5.10) has been determined theoretically by several workers for the case of uniaxial tension,with values ranging from 1.22 [213] through 1.25 to 3 [214]. Following the analysis of Rice and Tracey [215] it can be shown that

$$\frac{\cap}{3} = 0.5585 \sinh\left[\frac{3\sigma_M}{2\sigma_E}\right]$$

$$+ 0.008 \left[\frac{-3\dot{\epsilon}_2}{\dot{\epsilon}_1 - \dot{\epsilon}_3}\right] \cosh\left[\frac{3\sigma_M}{2\sigma_E}\right] \qquad (5.11)$$

where σ_M is the mean stress, σ_E the von Mises equivalent stress, $\dot{\epsilon}_1$, $\dot{\epsilon}_2$ and $\dot{\epsilon}_3$ are the principal strain rates. Rice and Tracey showed that $\cap$ would be equal to 0.9 for uniaxial tension, 1.65 for plane strain and 1.92 for balanced biaxial tension in a perfectly plastic solid (m=0). Alternatively, Budianski et al. predicted that $\cap$ would take different values depending on the magnitude of the strain rate sensitivity [216]. For example, in balanced biaxial tension, which would arise in the case of blowing a hemisphere from a circular

disc of material, Budianski et al. calculated that
$\cap$ would equal 1.75 when m=0.5; 1.94 when m = 0.33 and
2.64 when m = 0. Stowell et al. [217] have also shown
that the value of $\cap$ is dependent on the strain rate
sensitivity and hence will vary with the strain
rate. The relationship given by Stowell et al. for
uniaxial tension can be readily extended to
multiaxial states of stress. Following the analysis
of Cocks and Ashby [202]   it can be shown that
[218]

$$\cap = \frac{3}{2} \left[\frac{m+1}{m} \right] \sinh \left[2 \left[\frac{2-m}{2+m} \right] \left[\frac{\sigma_M}{\sigma_E} \right] \right] \tag{5.12}$$

or [193,216]

$$\cap = \frac{3}{2} (1+0.932m - 0.432m^2)^{1/m} \left[\frac{\sigma_M}{\sigma_E} \right] \tag{5.13}$$

The term containing the ratio of the mean stress,
σ_M, to the von Mises equivalent stress, σ_E, in
equations (5.12) and (5.13) defines the
triaxiality of stress local to the grain
boundary. The form of this term is dependent on the
geometry of deformation and may be rewritten in
terms of a geometric constant, K_S, and the
superimposed hydrostatic pressure, P. Thus

$$\frac{\sigma_M}{\sigma_E} = \left[\frac{k_S}{3} - \frac{P}{\sigma_E} \right] \tag{5.14}$$

To establish the values of k_S and k_D
appropriate to superplastic flow, two modes of
deformation need to be considered. Firstly, the
case of rigid grains where there is no grain
boundary sliding and secondly, the case of freely
sliding grains, where the state of stress on the
grain boundaries is dependent on the extent of
sliding. As the voids are located on the boundary
they experience the local rather than the remote
state of stress and thus their rate of growth is,
to a large extent, dependent on the fraction of the
total strain carried by the grain boundaries.

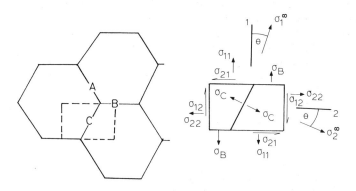

Figure 5.7 Geometry adopted by Cocks and Ashby [202] to determine the stress state local to a freely sliding grain boundary.

5.4.3 Determination of stress state local to the grain boundary

Consider first the two-dimensional arrangement of hexagonal grains shown in figure 5.7 (see also figure 4.7 [149]) [202]. The imposed stresses σ^∞_1 and σ^∞_2 are applied at some arbitrary angle θ with respect to the orthogonal axes defined by σ_{11} and σ_{22}. Resolving the forces due to the remotely imposed stresses and those acting on the continuum body, σ_{11} and σ_{22}, then

$$\sigma_{11} = \sigma^\infty_1 \cos^2\theta + \sigma^\infty_2 \sin^2\theta \qquad (5.15)$$

$$\sigma_{22} = \sigma^\infty_1 \sin^2\theta + \sigma^\infty_2 \cos^2\theta \qquad (5.16)$$

If the boundaries are freely sliding then the shear stresses will be fully relaxed, i.e. $\tau_a=\tau_b=\tau_c=0$ and the stress, σ_c, acting normal to the grain boundary will be [149,202]

$$\sigma_c = 1/2(\sigma^\infty_1 + \sigma^\infty_2) + (\sigma^\infty_1 - \sigma^\infty_2)\cos2\theta \qquad (5.17)$$

The stress on the grain boundary will vary in a periodic pattern with a repeat every 60°. The average stress on the boundary is obtained by integrating equation (5.17) over the range 0 to $\pi/6$ (30°), since there is an equal probability of the hexagonal grains being present in either the alternative 'flat' orientation or in the upright orientation.

$$\bar{\sigma}_c = \frac{\int_0^{\pi/6} \sigma_c \, d\theta}{\int_0^{\pi/6} d\theta} \tag{5.18}$$

$$\bar{\sigma}_c = \sigma^\infty_1 \left[\frac{1}{2} + \frac{3\sqrt{3}}{2\pi} \right] + \sigma^\infty_2 \left[\frac{1}{2} - \frac{3\sqrt{3}}{2\pi} \right] \tag{5.19}$$

$$\bar{\sigma}_c \approx 4\sigma^\infty_1/3 - \sigma^\infty_2/3 \tag{5.20}$$

However, in a real material grain motion will occur in three dimensions not two. We now assume that there is an equal probability that the sliding boundary will lie in the plane containing σ_{11} and σ_{33}, as well as in the plane containing σ_{11} and σ_{22}. For sliding in three dimensions, the average stress acting normal to the sliding boundary, σ_1, is now taken as the average of the two orientations, i.e.

$$\sigma_1 = 4/3 \; \sigma^\infty_1 - 1/6 \; (\sigma^\infty_2 + \sigma^\infty_3) \tag{5.21}$$

Normalising with respect to the von Mises equivalent stress we obtain

$$\frac{\sigma_1}{\sigma_E} = \frac{1}{2} \left[\frac{3\sigma^\infty_1}{\sigma_E} - \frac{\sigma^\infty_M}{\sigma_E} \right] \tag{5.22}$$

In order to obtain the local mean stress, σ_m, it is now assumed that the local deformation rate is the same as the remote deformation rate, i.e. the deviatoric component (S) of the local and remote stress fields is the same. As

$$S = \frac{2\sigma_1}{3} - \frac{1}{3}(\sigma_2 + \sigma_3) = (\sigma_1 - \sigma_M) \qquad (5.23)$$

then

$$\sigma_1 - \sigma_M = \sigma^\infty_1 - \sigma^\infty_M \qquad (5.24)$$

Substituting equation (5.21) for σ_1 and solving for σ_M

$$\sigma_M = \frac{3\sigma^\infty_1}{2} - \frac{\sigma^\infty_M}{2} - \sigma^\infty_1 + \sigma^\infty_M$$

$$\sigma_M = \frac{\sigma^\infty_1 + \sigma^\infty_M}{2} \qquad (5.25)$$

Now divide through by σ_E to obtain the local degree of triaxiality

$$\frac{\sigma_M}{\sigma_E} = \frac{1}{2}\left[\frac{\sigma^\infty_1}{\sigma_E} + \frac{\sigma^\infty_M}{\sigma_E}\right] \qquad (5.26)$$

Having defined both the remote and local extent of triaxiality in terms of the three remotely applied principal stresses ($\sigma^\infty_1, \sigma^\infty_2$ and σ^∞_3) it is now possible to obtain k_D and k_S for any deformation geometry.

5.4.4 Uniaxial tension with a superimposed gas pressure P.

The generalised form of the applied stress system for uniaxial deformation in the presence of a confining pressure, P, is given by

$$
\sigma_{ij} = \begin{bmatrix} \sigma - P & 0 & 0 \\ 0 & -P & 0 \\ 0 & 0 & -P \end{bmatrix} \tag{5.27}
$$

For the case of zero grain boundary sliding, the ratio of the maximum principal stress to the equivalent stress is

$$
\frac{\sigma^{\infty}_1}{\sigma_E} = \frac{\sigma_E - P}{\sigma_E} = 3/3 - P/\sigma_E \tag{5.28}
$$

and the ratio of the mean stress to the equivalent stress is

$$
\frac{\sigma^{\infty}_M}{\sigma_E} = \frac{\sigma_E - P - P - P}{3\sigma_E} = 1/3 - P/\sigma_E \tag{5.29}
$$

From eqns (5.22) and (5.28) it is possible to obtain the ratio of the maximum principal stress to the equivalent stress for the case of freely sliding grains;

$$
\sigma_1/\sigma_E = 1/2 \ [3(\sigma_E - P)/\sigma_E - (1/3 - P/\sigma_E)]
$$

$$
= 1/2 \ (8/3 - 2P/\sigma_E)
$$

$$
\sigma_1/\sigma_E = 4/3 - P/\sigma_E \tag{5.30}
$$

and from eqns (5.26) and (5.29), the ratio of the mean stress local to the sliding grain boundaries to the equivalent stress;

$$
\sigma_M/\sigma_E = 1/2 \ [(\sigma_E - P)/\sigma_E + (1/3 - P/\sigma_E)]
$$

$$
= 1/2 \ (4/3 - 2P/\sigma_E)
$$

$$
\sigma_M/\sigma_E = 2/3 - P/\sigma_E \tag{5.31}
$$

Comparing equations (5.28) to (5.31) with equations (5.9) and (5.14), then from the preceding analysis, values of $3 \leq k_D \leq 4$ and $1 \leq k_S \leq 2$ for uniaxial deformation are obtained. However, superplastic bulge forming involves the deformation of sheet material into two-dimensional shapes so it is also necessary to consider two additional geometries, balanced biaxial and plane strain deformation.

5.4.5 Balanced biaxial deformation

Consider the forming of a hemispherical section under a forward forming pressure $P+\delta P$ against a back pressure P. If the thickness of the sheet, h, being formed is uniform and the radius of curvature is ρ then the generalised state of stress is defined by

$$\sigma_{ij} = \begin{bmatrix} \delta P \rho/2h - P & 0 & 0 \\ 0 & \delta P \rho/2h - P & 0 \\ 0 & 0 & -(P + \delta P/2) \end{bmatrix} \quad (5.32)$$

For the majority of practical forming operations $\rho \gg h$; hence, the von Mises equivalent stress is approximately $\delta P \rho/2h$. It then follows that for zero grain boundary sliding

$$\sigma\infty_1/\sigma_E = 3/3 - P/\sigma_E \quad (5.33)$$

and

$$\sigma\infty_M/\sigma_E = 2/3 - P/\sigma_E \quad (5.34)$$

For the case of freely sliding grain boundaries

$$\sigma_1/\sigma_E = 3.5/3 - P/\sigma_E \quad (5.35)$$

and

$$\sigma_M/\sigma_E = 2.5/3 - P/\sigma_E \quad (5.36)$$

Hence for balanced biaxial forming, $3 \leq k_D \leq 3.5$ and $2 \leq k_S \leq 2.5$.

5.4.6 Plane strain forming

Consider the bulge forming of a rectangular pan where the length of the pan is much greater than its width. Under such conditions deformation parallel to the length of the pan is negligible and can therefore be approximated to plane strain. If the forward forming pressure is $P+\delta P$ and the back pressure is P, then the state of stress within a sheet of thickness h and curvature ρ is defined by

$$
\sigma_{ij} = \begin{bmatrix} \delta P\rho/h - P & 0 & 0 \\ 0 & \delta P/2(\rho/h-1/2) - P & 0 \\ 0 & 0 & -(P + \delta P/2) \end{bmatrix} \quad (5.37)
$$

The von Mises equivalent stress is approximately $\sqrt{3}\delta P\rho/4h$ since $\rho \gg h$. It then follows that for zero grain boundary sliding, the ratio of the remote maximum principal stress to the equivalent stress and the ratio of the remote mean stress to the equivalent stress, is given by

$$\sigma\infty_1/\sigma_E = 2\sqrt{3}/3 - P/\sigma_E \quad (5.38)$$

and

$$\sigma\infty_M/\sigma_E = \sqrt{3}/3 - P/\sigma_E \quad (5.39)$$

respectively, while for the case of freely sliding grain boundaries we find

$$\sigma_1/\sigma_E = 2.5\sqrt{3}/3 - P/\sigma_E \quad (5.40)$$

and

$$\sigma_M/\sigma_E = 1.5\sqrt{3}/3 - P/\sigma_E \quad (5.41)$$

Hence, for plane strain forming, $2\sqrt{3} \leq k_D \leq 2.5\sqrt{3}$ and $\sqrt{3} \leq k_S \leq 1.5\sqrt{3}$.

124

The relationships for the maximum principal stress and the mean stress for each deformation geometry are summarised in Table 5.1 .

In general, the value of both k_D and k_S, for a particular geometry, will lie between the limiting values derived above since only a fraction of the total strain accrued during superplastic flow is a direct consequence of grain boundary sliding. To a first approximation, the value of k_D or k_S can be obtained by linear interpolation between the two extremes cited above. The limiting values of k_D (for diffusive growth) and k_S (for strain controlled growth) are summarised in Table 5.2.

5.5 Analysis of Cavity Growth

When the rate of change of void radius per unit strain predicted by each of the three possible void growth mechanisms (eqns (5.7), (5.8) and (5.10) is calculated, then for the majority of superplastic materials which are deformed under conditions where optimum superplasticity is observed, the growth rate of an individual void is dominated in the first instance by diffusion (eqn (5.7)). However, once the void attains a radius between 0.5 and 1.5 μm then strain controlled growth (eqn (5.10)) dominates. Only in materials with a very fine grain size in which the optimum strain rate for superplasticity is $<10^{-4}$/s does superplastic diffusional growth (eqn (5.8)) become significant.

Each of the three equations describing the rate of void growth per unit strain will predict a different variation of void volume with strain. Providing that there is little void nucleation or coalescence during deformation then the void volume fraction, C_v, will be an integer multiple of the volume, V, of an isolated void and the two are therefore equivalent. A plot of the measured void volume fraction with strain should then enable the dominant void growth mechanism to be identified [196]. For the case of conventional diffusion controlled growth the void volume fraction should increase linearly with strain:

$$dr/d\epsilon \ \alpha \ 1/r^2 \qquad \text{hence} \qquad C_v \ \alpha \ \epsilon \qquad (5.42)$$

Table 5.1 Summary of the pressure dependencies (approximate form) of the stresses local to a rigid grain boundary and to a freely sliding grain boundary as derived from the analyses of Cocks and Ashby [202] and of Beere [149].

Model	Stress	Uniaxial	Biaxial	Plane Strain
Diffusion controlled cavity growth				
No sliding	$\sigma^{\infty}_1/\sigma_E$	$1 - P/\sigma_E$	$1 - P/\sigma_E$	$2/\sqrt{3} - P/\sigma_E$
Freely sliding	σ_1/σ_E	$4/3 - P/\sigma_E$	$7/6 - P/\sigma_E$	$5/2\sqrt{3} - P/\sigma_E$
Strain controlled cavity growth				
No sliding	$\sigma^{\infty}_M/\sigma_E$	$1/3 - P/\sigma_E$	$2/3 - P/\sigma_E$	$1/\sqrt{3} - P/\sigma_E$
Freely sliding	σ_M/σ_E	$2/3 - P/\sigma_E$	$5/6 - P/\sigma_E$	$3/2\sqrt{3} - P/\sigma_E$

Table 5.2 Summary of the limiting values of the geometric constant for cavity growth for diffusive and strain controlled growth and for no grain boundary sliding (x = 0) and 100% grain boundary sliding (x = 1)

Deformation mode	k_D		k_S	
	x=0	x=1	x=0	x=1
Uniaxial	3	4	1	2
Balanced biaxial	3	3.5	2	2.5
Plane strain	$2\sqrt{3}$	$2.5\sqrt{3}$	$\sqrt{3}$	$1.5\sqrt{3}$

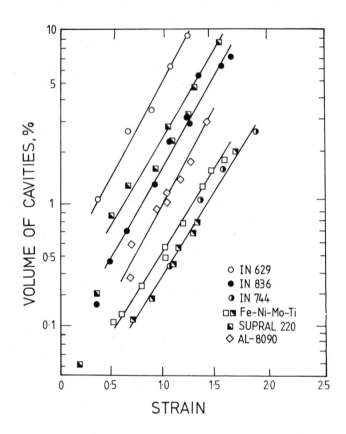

Figure 5.8 Experimentally determined relationship between void volume fraction and strain for several superplastic alloys deformed within Region II.

If the voids are larger than the grain size and diffusional growth continues to dominate, then the void volume fraction will increase as the cube root of strain:

$$dr/d\epsilon = \text{constant} \qquad \text{hence} \qquad C_v \propto \epsilon^{1/3} \qquad (5.43)$$

If void growth is strain controlled then the void volume fraction would be expected to increase exponentially with strain:

$$dr/d\epsilon \propto r \qquad \text{hence} \qquad C_v \propto e^{\epsilon} \qquad (5.44)$$

$$\text{or} \qquad C_v = C_o\exp(\cap\epsilon) \qquad (5.45)$$

The variation of the volume fraction of voids with strain has been measured for several superplastic alloys including α/β copper alloys [105,106,217, 219-221], low alloy steels [63], α/τ stainless steels [66,222] and several aluminium alloys [198,218,223,224]. In all cases the variation of the void volume fraction with strain is best described by an exponential relationship indicating that for the most part void growth is plasticity controlled (Fig. 5.8). During superplastic flow, void growth involving plasticity control would not occur in the manner assumed by Hancock [214] or Tait and Taplin [225] where the individual voids, with radii very much greater than the grain size, grew in an essentially isotropic matrix. In practice, where the dimensions of the grains and cavities are comparable, void growth is topological in nature and involves displacements of adjoining grains relative to each other by grain boundary sliding. This results in irregular shaped voids while the grains themselves remain equiaxed. The most significant deviations from strain controlled growth occur at low void volume fractions and low strains. Under these conditions the voids are small and isolated and diffusional growth would be expected to dominate. Unfortunately, few measurements of the variation of void volume fraction with strain exist for low strains or for deformation conditions outside the normal superplastic regime.

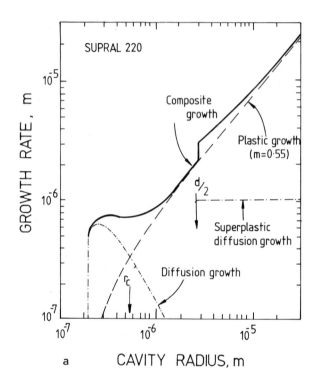

Figure 5.9 (a) Calculated variation of cavity growth rate with strain for diffusive, plastic and combined mechanisms.

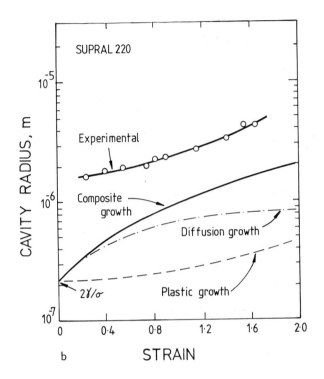

(b) Calculated variation of cavity radius with strain for each of the above mechanisms together with the experimental measurements. (Supral 220, $T = 460°C$, $\dot{\epsilon} = 1.2 \times 10^{-3}/s$, $\sigma_E = 10 MNm^{-2}$).

The change in void radius with strain and the variation of void growth rate (dr/dϵ) with void radius, calculated from eqns (5.7), (5.8) and (5.10), in which the effect of surface tension has been included, have been plotted in figure 5.9 using data for Supral 220 as an example (T = 460°C; $\dot{\epsilon} = 1.17 \times 10^{-3}$/s; $\sigma = 10$ Nmm^{-2}). In calculating the variation of the radius as a function of strain it was assumed that the void pre-existed and that at the start of deformation it had the minimum size consistent with thermodynamic stability ($r_o = 2\Gamma/\sigma$).

It can be seen from figure 5.9a that for voids with radii up to ~0.3 μm, the rate of void growth by diffusion is the greater and thus dominates. However, for void radii >0.3 μm and <1 μm both the plasticity and diffusion controlled processes proceed at a similar rate. For voids with radii in excess of 1 μm, growth occurs almost exclusively by the plasticity mechanism. It is evident from figure 5.9b that it would be impossible for the voids to reach the sizes that are observed experimentally if each growth process operates to the exclusion of the other. If the two processes act independently on the growing void and their effects are additive (composite growth law) then much larger void radii can be attained at a given strain. However, even after strains equivalent to 600% elongation, the largest voids would be expected to be only 4 μm in diameter. This size is significantly smaller than that observed experimentally (Fig. 5.10). In order that such large voids can be formed either the initial voids must be larger than assumed or substantial void coalescence must occur during deformation. Since voids are generally not resolvable in the undeformed material it must be concluded that void agglomeration is a key process in the development of cavitation damage.

5.6 Void Growth by Coalescence

Void growth during superplastic flow is, for the most part, strain controlled. Therefore, the rate of increase of the volume of an isolated void with strain will depend solely on its volume. When two

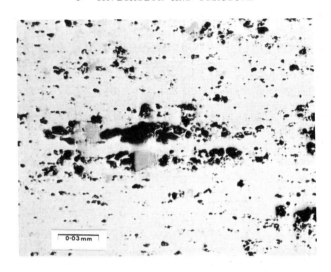

Figure 5.10 Voids observed at a strain of 1.6. (Supral 220, T = 460°C, $\dot{\epsilon}$ = 1.17 x 10^{-3}/s, σ_E = 10 MNm^{-2})

or more voids coalesce the void produced is larger than either of those from which it formed and, as such, its volume would increase at a more rapid rate than would that of either of its constituent voids [217]. However, the effect of a sudden increase in the growth rate of a number of isolated voids on the rate of increase of the total void volume fraction is offset by a decrease in the total number of voids. There should be <u>no net change</u> in the rate of increase in the total void volume fraction with strain. Void coalescence serves to extend the distribution of void sizes to larger radii, and hence to increase the mean void growth rate. Void coalescence therefore produces the large isolated voids which are most likely to have an adverse effect on the post-forming mechanical properties of superplastic alloys.

The extent of void coalescence during superplastic
flow is dependent on the size, number of voids per
unit density (population density) and the spatial
distribution of the voids within a material. The
effect of pairwise coalescence, in a system of
randomly distributed voids, on the mean void growth
rate can be estimated from [223]

$$\frac{d\bar{r}}{d\epsilon} = \frac{8C_v\Phi(\delta\epsilon) \cap (0.13r - 0.37(dr/d\epsilon)_i\delta\epsilon + (dr/d\epsilon)_i}{1 - 4C_v\Phi(\delta\epsilon) \cap \delta\epsilon}$$

(5.46)

where C_v is the instantaneous volume fraction of
voids, $\cap$ is defined in eqn (5.12), $\delta\epsilon$ is a small
increment of strain, $(dr/d\epsilon)_i$ is the rate of void
growth per unit strain of an isolated void and
$\Phi(\delta\epsilon)$ is given by

$$\Phi(\delta\epsilon) = (1 + \cap\delta\epsilon/3 + (\cap\delta\epsilon)^2/27)$$ (5.47)

At low volume fractions of voids ($C_v < 0.01$) void
coalescence has a negligible effect on the mean
void growth rate, but becomes much more significant
at volume fractions greater than ~0.05.

Unfortunately, it is not possible to calculate the
variation of the volume fraction of voids with
strain without a quantitative understanding of the
functions describing the distribution of void
sizes, the overall population density of voids and
their variation (if any) with strain, i.e. a
quantification of void nucleation. However, by
combining experimental observations of the size
distributions and population densities of voids in
superplastically deformed alloys, a
semiquantitative analysis of the evolution of
cavitation damage, which can include the effects of
superimposed hydrostatic pressure, can be made.

Measurements of the void size distributions using
optical microscopy have shown that the number of
voids per unit volume in both copper and aluminium
base alloys, is between 10^{14} and $5 \times 10^{15}/m^3$

134

after deformation to various strains under a wide
range of strain rates and temperatures. The
relative independence of the total void population
on the deformation parameters could merely reflect
the limited resolution of optical measurments.
However, it could also be inferred that the
majority of the voids pre-existed and that
nucleation during superplastic flow was minimal.

5.7 Role of Hydrostatic Pressure on Cavity Growth

Experimental results have shown that:

1. Cavities formed during superplastic flow can be
   removed by subsequent compressive flow [197].
2. Annealing of cavitated material after forming
   reduces the volume fraction of voids [226].
3. Post-forming hot isostatic pressing has the
   potential to remove completely any voids formed
   during superplastic forming [199,266].

Unfortunately, the annealing of superplastically
formed parts is only capable of removing the very
small voids, where surface tension is capable of
driving the diffusion process necessary to sinter
the voids within a relatively short period of time.
Similarly, the hipping process may be limited in
application in view of its cost, its restriction on
component size and the possiblity that the voids
will reappear during any subsequent heat treatment
[227].

From an examination of the equations describing
cavity growth presented in sections 5.4.1 and
5.4.2, it is apparent that the superimposition of a
confining gas pressure during superplastic
deformation should restrict the rate of cavity
growth since it reduces both the maximum principal
stress and the mean stress. The experimental
evidence cited in Section 5.5 supports the view
that strain rather than diffusion is the dominant
void growth mechanism during superplastic
deformation. A simple criterion for determining the
appropriate level of hydrostatic pressure is easily
derived from equation (5.14) namely

$$\sigma_M/\sigma_E \leq 0 \tag{5.48}$$

The only uncertainties that remain are in the establishment of the appropriate form of the equation to represent σ_M/σ_E (see Table 5.2) and of the magnitude of the flow stress of the material under investigation. Equation (5.10) then corresponds to the condition that $\cap$ takes a zero or negative value.

Quantitative investigations of the role of superimposed pressure on the accumulation of cavitation damage during superplastic flow are limited. The majority of the available data relates to commercial aluminium alloys such as Supral 220, Al-7475 and Al-8090 [198,218,223,228]. Bampton and co-workers [198,228] reported on the effects of hydrostatic pressure on cavitation in Al-7475 deformed in uniaxial, balanced biaxial and plane strain conditions. Similar results have been obtained in subsequent studies on Al-7475, Al-8090 and Supral 220 in both uniaxial and balanced biaxial deformation [218] (Fig. 5.11). In general, increasing the superimposed pressure was found to:

1. Decrease the rate at which the volume fraction of voids increased with strain.
2. Decrease the level of cavitation at a given strain
3. Displace to higher strains the strain required prior to the detection of cavitation.
4. Increase, to a limiting value, the strain to failure.

The level of cavitation at a given strain has been found by experiment to be governed primarily by the ratio of the mean stress to the flow stress, (σ_M/σ_E), rather than the ratio of the maximum principal stress to the flow stress, (σ_1/σ_E). This observation is in agreement with the theoretical predictions of Section 5.5. It was also noted that the level of cavitation fell markedly when the ratio of the mean stress to the flow stress was less than or equal to zero. The latter observation is consistent with the plasticity or strain dominated model of void growth, and supports the use of equations (5.12) or (5.13) to define $\cap$

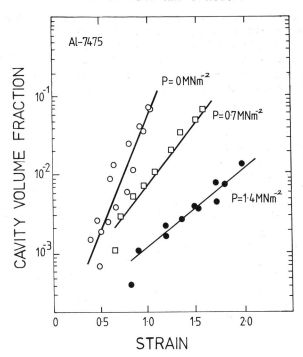

Figure 5.11 The variation of void volume fraction with strain in Al-7475 at different superimposed pressures for deformation in balanced biaxial tension [218].

since the cavity growth rate factor is zero when σ_M/σ_E is zero (Fig. 5.12).

It has been found for a number of superplastic aluminium alloys that the cavity growth rate factor, η, is strongly dependent on the applied state of stress with cavitation being virtually eliminated for superimposed pressures of approximately one half of the uniaxial flow stress (Fig. 5.12) [218]. Again, the variation of the volume fraction of voids with strain under

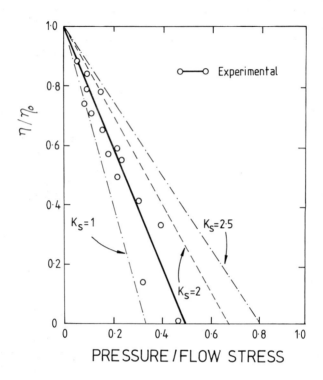

Figure 5.12 The variation of the normalised cavity growth rate parameter (η/η_o) with pressure normalised with respect to the flow stress, as calculated for different values of the geometric factor K_s. Values determined experimentally for several aluminium alloys deformed in uniaxial tension are also shown (o).

different confining pressures was found experimentally to be consistent with a strain controlled model of void growth. By far the most significant factor in the void growth process, and hence in determining the level of cavitation that is observed at a given strain, strain rate and temperature, is the ratio of the superimposed pressure to the flow stress, P/σ_E.

In order to calculate the effect of superimposed pressure on the rate of accumulation of cavitation damage it is necessary to estimate the number of voids in unit volume. If it is assumed that ~5×10^{14} voids/m³ pre-exist, and that the voids are the minimum size consistent with thermodynamic stability at the onset of deformation (i.e. $r_o = 2\Gamma/(\sigma-P)$), then it is possible to calculate the volume fraction of voids at a given strain for any level of superimposed pressure. In figure 5.13, the variation of the volume fraction of voids with strain is plotted for Supral 220 deformed at a strain rate and temperature where the strain rate sensitivity, m, is at a maximum. For the case of P=0, the volume fraction of voids rises rapidly with strain, reaching 1% at an elongation of 250% and ~8% at an elongation of 500%. In contrast, the application of a hydrostatic pressure equal to half the flow stress, $P = 5 \, N \, mm^{-2}$, limits void growth to such an extent that the volume fraction of voids would be less that 0.1% at an elongation of 650%.

Metallographic studies have been made of the void size distributions and void morphologies which develop during the forming of aluminium alloys at various levels of superimposed pressure. The work has shown that the number of voids per unit volume is little affected by the pressure although the size to which they grow decreases dramatically with increasing pressure (Fig. 5.14). The restriction on void growth imposed by the confining pressure is coupled with a change in void morphology. At zero or low confining pressures the voids are normally observed to have grown in a cracklike manner. Voids which were initially spherical spread along the grain boundary becoming elongated, often with lengths far greater than their height. The change

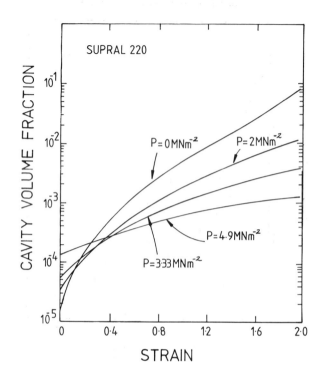

Figure 5.13 Calculated variation of void volume
fraction with strain for different superimposed
pressures. (Supral 220, T = 460°C, $\dot{E}$ = 1.17 x 10^{-3}/s,
σ_E = 10 MNm^{-2})

in void morphology from spherical to cracklike is
a consequence of the inability of surface diffusion
to redistribute matter from the poles of the void
(low curvature) to the void neck (high curvature),
at a rate commensurate with that at which atoms
leave the neck along the grain boundary [229-230].

The spreading of a void parallel to the grain
boundary allows the rate of increase of its major
axis to be large, while its rate of change of volume

140

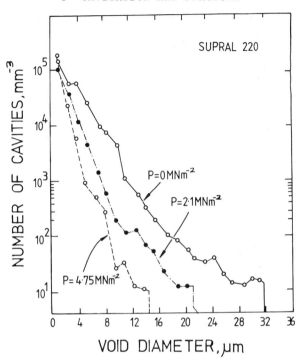

Figure 5.14 Measured void size distributions of Supral 220 at 450% elongation with varying superimposed pressures.

is still that predicted by eqn (5.7) [230]. Thus at low confining pressures, void coalescence occurs frequently and large voids quickly develop. Once formed, the larger voids grow as a consequence of the plastic deformation of the surrounding matrix, rather than by diffusion, and the growth rate increases dramatically. However, at higher confining pressures, the voids remain spherical as the driving force for boundary diffusion is reduced. Coalescence is therefore limited, and the large void

Table 5.3 The effect of pressure on the extent of cavitation in one batch of Al-8090 deformed at 520°C and $10^{-3}/s$.

Strain (%)	Pressure (Nmm^{-2})				
	0	0.7	1.4	2.1	3.05
200	1.6%	0.25%	0.2%	0.08%	n/d
300	5.0%	0.5%	0.35%	0.1%	n/d
400	failed	1.1%	0.55%	0.11%	0.02%
500	failed	failed	failed	0.13%	0.02%
600	failed	failed	failed	failed	0.12%

(n/d - no cavitation detected by densitometry)

networks which are observed at lower confining pressures are prevented from developing.

Despite the low flow stresses of the majority of commercially useful superplastic alloys ($\sigma < 10$ Nmm^{-2}) it would be both costly and difficult to apply confining gas pressures in excess of the flow stress to eliminate cavitation completely. Therefore the use of hydrostatic pressure in superplastic forming has been to limit the development of cavitation to such an extent that those voids which do form ($C_v < 0.1\%$) would be relatively innocuous, being small, spherical and essentially isolated. If it is assumed that ~50% of the total strain is attributable directly to grain boundary sliding, then the minimum hydrostatic pressure that would be required to eliminate cavitation would be

$$P \geq 0.75 \, \sigma(\epsilon) \tag{5.49}$$

for balanced bi-axial tension, and

$$P \geq 0.72 \, \sigma(\epsilon) \tag{5.50}$$

for plane strain deformation, where $\sigma(\epsilon)$ is the measured uni-axial flow stress at the forming strain rate at the point of greatest strain. (The values quoted are midway between those for no sliding and those predicted by the models of Cocks and Ashby, and Beere (Tables 5.1 and 5.2). In practice, confining pressures would not normally exceed 3.5 Nmm^{-2} (500 psi), and thus the use of a confining pressure is limited to the slow strain rate forming of alloys such as Al-7475 and Al-8090 whose flow stresses at strain rates of 2×10^{-4} to 5×10^{-4}/s are of the order of 3-7Nmm^{-2}.

The effectiveness of back pressure in reducing the level of cavitation in superplastically formed components, and of increasing the forming potential of marginally superplastic but high strength alloys is readily apparent form Table 5.3. It can be seen that the elongation to failure is increased from ~300% to >600% while the level of cavitation is reduced from 5% by volume at zero

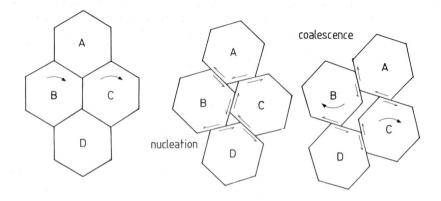

Figure 5.15 Schematic illustration of void
nucleation as a result of unaccommodated grain
rotation.

superimposed pressure to nil at an elongation of
300% with a superimposed pressure of 3.05 N mm⁻²
(450 psi).

5.8 Cavity Nucleation

5.8.1 Introduction

Strain is accumulated during superplastic
deformation primarily as a result of grain boundary
sliding, rather than by the elongation of the
grains themselves. The deformation occurs
topologically and the formation of intergranular
voids is therefore intimately linked to the
mechanisms of deformation. If rigid grains are to
move relative to each other then the formation of
voids becomes geometrically necessary [126,132,231]
(Fig. 5.15). Fortunately, however, the rotational
and translational displacements of the grains can
be accommodated by the redistribution of matter
within a narrow zone (mantle) around the grain
boundaries, with the interior of the grains
remaining essentially undeformed [131]. It has been
proposed that the width of such a mantle needs only

be of the order of 0.07 times the grain diameter. In fine grain superplastic materials this is a very small width indeed (30-70 nm). The accommodation of grain boundary sliding, as has been shown, can be achieved in several ways [128]:

1. Grain boundary and volume diffusion.
2. Glide and climb of grain boundary dislocations.
3. Glide and climb of lattice dislocations both across and/or around the grains.

The driving force for the above processes is provided by the variation, on a local scale, in the magnitude of the unrelaxed grain boundary normal tractions and shear stresses [134,232]. The presence of microstructural defects such as grain boundary ledges, triple points and intergranular particles, can contribute substantially to raising the stress level local to the sliding grain boundaries. When the kinetics of the accommodating processes fail to match the requirements imposed by the deformation rate then the stresses do not relax sufficiently quickly and cavities may nucleate.

The factors which influence cavity nucleation are those which relate to the microstructure of the alloy such as grain size, the type, volume fraction and distribution of hard particles, the proportions and physical properties of the major phases, and those which are associated with the deformation conditions, for example, strain, strain rate, temperature and stress state. It will become evident however, that there is a strong interrelationship between the microstructural and deformation factors. An expression has been proposed [195] in which the strain rate, $\dot{\epsilon}_c$, below which void nucleation at a grain boundary particle of diameter D, is likely to be inhibited by diffusional stress relaxation can be predicted. This is given by

$$\dot{\epsilon}_c = \frac{11.5\sigma\Omega}{xdD^2} \frac{D_{gb}\delta}{kT} \qquad (5.51)$$

where d is the grain size, x the fraction of the
total strain carried by the grain boundaries, D_{gb}
the grain boundary diffusion coefficient, δ the
grain boundary width, and σ the flow stress at the
imposed strain rate. Hence, if the critical strain
rate was greater than the imposed strain rate,
cavity nucleation would be unlikely. It is clear
from the above relationship, which provides a
rationalisation of the interaction between the many
microstructural and mechanical aspects of
superplastic flow, that deformation at higher
temperatures and/or lower strain rates would
minimise cavitation. However, the latter conditions are
not commercially attractive since forming times
would be long and energy costs high. Furthermore,
structural instability at elevated temperatures can
result in substantial grain growth and particle
coarsening and a loss in superplasticity and/or
enhanced cavitation.

The minimum cavity radius which will be stable
under an applied tensile stress σ, in the presence
of a superimposed pressure, P, is given by

$$r \geq 2\Gamma/(\sigma-P) \tag{5.52}$$

where Γ is the surface energy. If the maximum
principal stress is similar in magnitude to the
flow stress, then for the majority of superplastic
materials the critical cavity radius would be
approximately 100 nm, or less if nucleation
occurred at a stress concentration. It is unlikely
that voids would nucleate spontaneously by vacancy
condensation, as the number of vacancies required to
form a stable void would be too large.

Approximately 2×10^8 vacancies would be required
to form a spherical void of radius 100 nm. If the
grain size of the material is typical of that found
in superplastic materials, i.e. 5 to 10 μm, then a
vacancy concentration of $\sim 10^{-5}$ would be
necessary if all the vacancies in the grain were to
condense simultaneously to nucleate a single void.
(By way of contrast, the vacancy concentration at
$0.8T_m$ is of the order of 10^{-7} to 10^{-6} even
when the effect of the applied stress is taken

into account). However, it has been suggested that grain boundary sliding does not occur at a uniform rate but in rapid bursts and on a very localised scale. Thus prior to sliding, the transient stresses generated could be sufficiently high to nucleate a very small void from a highly localised supersaturation of vacancies [134]. Once the stresses were relaxed by sliding the void would be unstable, but would require a finite time to decay. It is possible that the nucleated void could be stabilised by further sliding [229]. However, in practice it is likely that the majority of the cavities pre-exist, and are nucleated at second phase particles during the intense deformation associated with thermomechanical processing.

The majority of studies on cavitation in superplastic alloys have shown that the number of voids observed by optical microscopy increases with increasing strain. However, it should be recognised that optical microscopy has only a limited resolution, and that during the initial stages of deformation the smaller voids will not be resolved, even though they are present. Thus the number of voids will appear to increase with increasing strain. For example, a void will grow at a rate determined by eqn (5.10). Hence, for a material with a strain rate sensitivity equal to 0.5, a strain equal to ~350% elongation would be required before a void with an initial radius of 100nm became optically resolvable (the limit of resolution has been taken as 0.5 μm). It is therefore important, when making experimental measurements, to distinguish between true nucleation and the detection of voids which are already present in the material but which are initially unresolvable.

5.8.2 Hard Particles

Experimental investigations have shown a clear correlation between the presence of hard second phase particles and cavitation, while microduplex materials which are particle free do not readily cavitate. Such a correlation has been clearly demonstrated for the non-cavitating Pb-Sn eutectic. The addition of a third element leads to

147

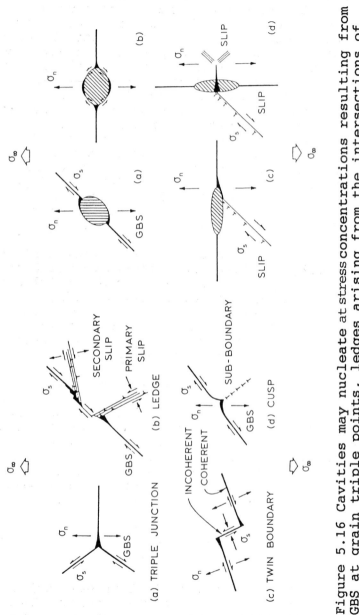

Figure 5.16 Cavities may nucleate at stress concentrations resulting from GBS at grain triple points, ledges arising from the intersections of slip bands or twins with the grain boundary; from stress concentrations adjacent to particles on sliding grain boundaries; and from dislocation pile-ups at the grain boundary particles.

the formation of intermetallic phases with varying
hardnesses, e.g. SbSn, Ag_3Sn and Cu_6Sn_5 [183].
On subsequent deformation, the cavities were
observed to have nucleated at the intermetallic
phase/matrix interfaces and this was attributed to
the limited ability of those phases to contribute
to the accommodation of grain boundary sliding. The
level of cavitation at a given strain was found to
increase, and the corresponding elongation to
failure decrease, as the hardness, volume fraction
and size of the intermetallic particles was
increased. The presence of large particles of
Ti(C,N) in α/τ steels [63], or of primary $ZrAl_3$
and $CuAl_2$ in Al-Cu-Zr alloys [40], of $Mg_3Cr_2Al_{18}$
and Fe and Si rich inclusions in Al-7475 [174], and of
cobalt silicide particles, CoSi and $CoSi_2$ in
CDA-638, appear to be largely responsible for
cavitation in these alloys (Fig. 5.16).

It is interesting to note that α/β titanium alloys
do not readily cavitate. The impurities which are
responsible for the formation of hard particles in
other materials, such as carbides, nitrides and
oxides in steels, and sulphides in copper base
alloys [105,220,233] are not as important in
titanium, as the impurity elements readily pass
into solution at elevated temperatures.

It has been shown that the particle/matrix
interfaces on sliding grain boundaries are the most
likely sites for voids to form [208], since stress
concentrations, poor bonding and a large radius of
curvature of the interface, combine to reduce
substantially both the critical void volume and
the energy barrier to nucleation (Fig. 5.16). The
steady state cavity nucleation rate, j, has been
predicted by Raj and Ashby [208,234] to be:

$$j = \frac{4\pi\Gamma}{\sigma_1\Omega} \frac{D_{gb}\delta}{\Omega^{1/3}} \; n.\exp\left[\frac{-4F_v\Gamma}{\sigma_1^2 kT}\right] \qquad (5.53)$$

where σ_1 is the maximum principal stress acting
on the particle, Γ the mean surface energy, F_v is
the shape factor which relates the volume of the
void to a characteristic dimension such as radius,

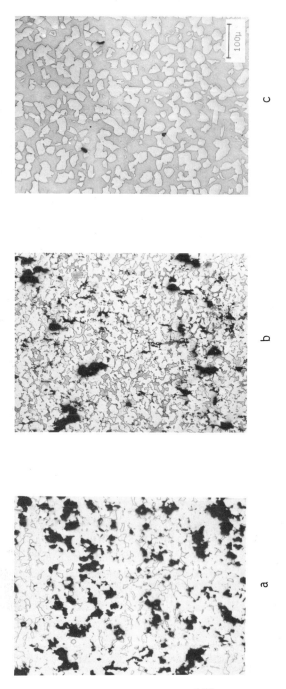

Figure 5.17a-c. The extent of cavitation damage immediately behind the fracture surface for different volume fractions of α phase in α/β Cu-Zn (a) 72%α, (b) 53%α and (c) 32%α.

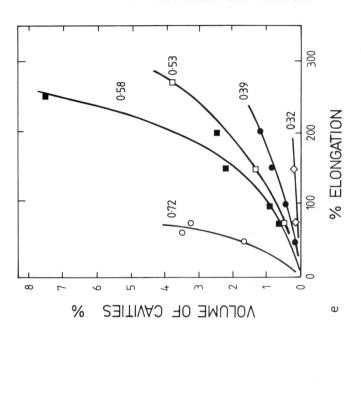

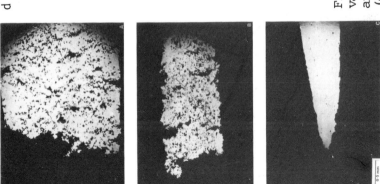

Figure 5.17d-e. (d) The appearance of the fracture surface also varies as the phase proportions are changed, the reduction in area increasing as the volume fraction of β phase increases. (e) Measured variation of the volume fraction of voids with strains for α/β Cu-Zn with varying α phase volume fractions.

and n is the density of nucleation sites on the grain boundary. The exponential term is the most dominant in eqn (5.53). This infers that nucleation would be more likely at higher stresses, lower surface energies, and when the void volume was small with respect to its effective radius.

In the absence of particles it has been argued that the stresses, (including transients at other stress concentrators such as triple points), and the necessary vacancy supersaturations are not attainable during deformation at elevated temperatures, and that conventional nucleation by vacancy condensation is impossible [235]. However, since cavities cannot be observed directly at the nucleation stage, the association of cavities with intermetallic or non-metallic particles can also be interpreted in terms of pre-existing defects such as regions of decohesion (embryonic cavities) which have developed during processing but which have a minimal volume at the commencement of deformation.

5.8.3 Phase Proportions and Characteristics

Studies of α/β brasses [17,221] have shown that the extent of cavitation during superplastic flow is strongly dependent on the relative proportions of each phase present in the microstructure, with cavitation decreasing as the volume fraction of β phase increases (Fig. 5.17). It can be seen in figure 5.17d that the cavitation behaviour also influences the fracture mode. The alloy with 72% α phase undergoes substantial cavitation and the coalescence and interlinkage of cavities leads to a pseudo-brittle fracture with a large cross-sectional area, whereas the alloy with 32% α phase, which shows little cavitation, pulls down to a fine point at failure. The alloy with approximately equal proportions of the two phases (53% α phase) shows an intermediate cross-sectional area at failure. The cavitation behaviour is consistent with the view that superplastic flow is accommodated primarily within the softer β phase [17,236]; the hard α phase acts in the same way as the intermetallic phases discussed in section 5.8.3. Shang and Suery have satisfactorily predicted cavity evolution in α/β

copper alloys in terms of the volume fraction of α phase [237]. It was proposed that the volume fraction of cavities which developed during deformation could be equated to the evolution of a second 'soft' phase (or an equivalent reduction in the volume fraction of hard α phase) which would further accommodate deformation, thus reducing the subsequent cavitation rate.

The α/β titanium alloys such as Ti-6Al-4V are structurally analogous to the α/β brasses, yet are remarkably resistant to cavitation. At the optimum deformation temperature the alloys usually contain ~40% by volume of β phase which accommodates grain boundary sliding and rotation, as the diffusion rate in the β phase is substantially greater than that in the α phase. At lower temperatures, the volume fraction of the β phase is reduced and cavitation is observed. Even at the optimum deformation temperature, rafting of the α grains into bands can result in cavitation since the local volume fraction of β is <40% [187]. However, in α/β titanium alloys the volume of cavites is usually minimal. At higher temperatures, the reduced volume fraction of α phase is less effective in stabilising the β grain size and the material ceases to be superplastic.

5.8.4 Grain Size

Experimental evidence is available that supports a correlation between increased cavity nucleation and larger grain sizes. In the aluminium base alloy Supral 220, where the volumetric growth rate of the cavities is almost independent of strain rate within the superplastic regime [238], the volume fraction of voids was found to increase with an increase in the grain size (Fig. 5.18). Moreover, grain growth during superplastic flow has been invoked as the reason for the 'apparent' continuous nucleation of cavities in both Al-7475 and Supral 220 [238,239]. In the highly superplastic Zn-22Al alloy, cavitation was found to be minimal when the grain size was less than 5 μm but increased quite markedly with initial grain sizes in excess of 5 μm [191].

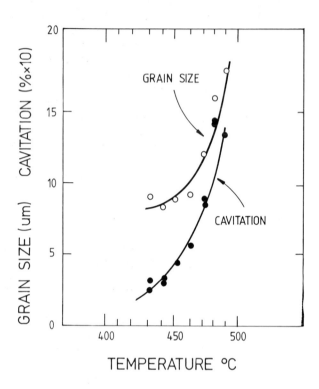

Figure 5.18 Measured variation of the volume fraction of voids and the grain size after deformation to 300% elongation at various temperatures within the superplastic regime (Supral 220)

The importance of microstructural stability in relation to cavitation is readily apparent from studies on Fe-C alloys. Plain carbon eutectoid steels with a fine grain ferrite/cementite structure show extensive cavitation at the α/Fe$_3$C interfaces [33]. During superplastic deformation, there is considerable coarsening of the cementite particles and this is accompanied by an increase in the ferrite grain size. By increasing the carbon content of the steel to 1.6 wt%, and hence markedly increasing the volume fraction of cementite, a very fine grain size can be stabilised. The structure can be further stabilised by the addition of ~1.5 wt% Cr which partitions to the cementite and increases its resistance to coarsening. The microstructure of the high carbon alloy is extremely resistant to cavitation, despite the increased volume fraction of cementite particles.

5.8.5 Strain Rate and Temperature

It might reasonably be expected that increasing the deformation temperature and/or reducing the strain rate would reduce cavitation. The flow stresses would generally be lower at the slower strain rates and higher temperatures, while there would be either a greater period of time to relax the stresses generated by grain boundary sliding at the lower strain rates, or the diffusion processes would become more rapid at the higher temperatures. However, the effect on cavitation of changing the strain rate or temperature during superplastic deformation varies widely depending on the material. The differences which arise are often the result of changes in the microstructure of the material, particularly the grain size.

In the aluminium alloy 7475, increasing the temperature and decreasing the strain rate results in a decrease in the overall level of cavitation at a given strain. In Supral 220, decreasing the strain rate from 10^{-2}/s to 10^{-3}/s results in a decrease in the overall level of cavitation at a given strain, but further reductions in the strain rate toward 10^{-5}/s result in a steadily

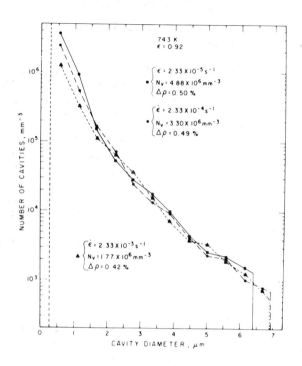

Figure 5.19 Measured void size distributions in a Ni-silver as a function of strain rate at 743K for a strain of 0.92.

increasing level of cavitation. It is unlikely that the increased levels of cavitation at the lower strain rates result from increased cavity growth since the value of the strain rate sensitivity and hence η, the cavity growth rate factor, is virtually constant over the range of strain rates from 10^{-5} to 10^{-2}/s [48]. At the lower strain rates, grain growth during superplastic flow is extensive and it would appear that the detrimental effect of increasing grain size is more than sufficient to offset the beneficial effect of reducing the strain rate. The same effect is observed on increasing the deformation temperature.

Studies of cavitation in α/β nickel silvers [105,220,233] and in α/τ stainless steels [240] have shown that both the volume fractions and size distributions of the cavities are independent of strain rate and temperature, and dependent only on strain (Fig. 5.19). As was shown earlier, void growth during superplasticity is strain controlled and, as such, the observed independence of the level of cavitation on both strain rate and temperature might infer a constant number of cavity nuclei; these observations could be interpreted as support for the view that either the majority of voids pre-exist, or inclusion particles dominate void nucleation in these systems.

5.9 Summary

It is clear that there is a strong interrelationship between the occurrence of cavitation, the microstructure of the superplastic material and the conditions under which it is deformed. There is also a large body of evidence to support the opinion that many of the cavities which become apparent during superplastic flow pre-exist, and are a natural consequence of the thermomechanical processing which has to be applied to the material in order to generate the initial fine grain microstructure.

However, the presence of numerous potential cavity nucleii need not be considered detrimental since a clear understanding of the cavity growth process has enabled the deformation conditions which are necessary to prevent cavity growth to be identified. The extent to which cavity growth occurs is controlled primarily by the strain imparted to the material. The application of a hydrostatic pressure of 0.5 to 0.8 times the uniaxial flow stress at the required deformation strain rate and temperature is both predicted and found to be sufficient to virtually eliminate cavitation from those systems in which it is normally apparent. More importantly, since the flow stress of superplastic materials is generally very low, the use of hydrostatic pressure in commercial forming operations becomes an extremely viable method of improving post-forming mechanical properties.

6
SUPERPLASTIC FORMING AND DIFFUSION BONDING

6.1 Introduction

There is considerable interest in the application of superplasticity, particularly in the aerospace industry. Unlike conventional materials, superplastics are extremely resistant to neck formation when deformed in tension, and this enables large uniform plastic strains to be attained without failure. This behaviour, as has been shown in the previous chapters, is derived from the high strain rate sensitivity of the flow stress. The high strain rate sensitivity is developed at commercially viable strain rates (5×10^{-4} to $10^{-2}/s$), while the stresses required to attain such strain rates are generally very low. The exploitation of superplasticity is, however, limited by several factors. Firstly, the reluctance of designers to adopt more novel forming procedures and new materials; secondly, the need for careful process control to confine forming to the rather narrow range of temperatures and strain rates over which a given material exhibits a high degree of superplasticity, and thirdly, the relatively small number of alloys presently available that can be both processed to give a fine stable grain size and that will also be capable of attaining the required strength, fatigue resistance and toughness, in the as-formed condition.

The bulk of commercial superplastic forming has been carried out on aluminium alloys such as

Supral 100, 150, 220 and 5000, Al-7475 and Lital A (Al-8090), and on titanium alloys such as Ti-6Al-4V, (see Tables 2.1 and 2.3). The aluminium alloys are usually formed at temperatures between 430 and 540°C depending on the composition, while the titanium alloys require temperatures between 860 and 920°C. The strain rates employed lie between 2×10^{-4} and 10^{-3}/s, though in the case of the Supral group of alloys strain rates approaching 10^{-2}/s may be used

The flow stresses of superplastic materials, which generally lie between 2 and 10 MN m^{-2}, enable gas pressure to be used to drive deformation rather than liquid hydraulic or mechanically applied loads. Superplastic sheet forming processes are therefore fundamentally different from the processes that are used in conventional metal forming, and have more in common with the techniques used for shaping thermoplastics.

6.2 Forming processes

6.2.1 Simple female forming

The most simple form of superplastic forming involves the free bulging of a sheet, which is clamped around its periphery, into a female mould (Fig 6.1 (1)-(3)). During the initial stages of deformation, when the sheet is not in contact with the tool, deformation is concentrated at the pole of the dome and consequently this region shows the greatest strain. Once the pole comes into contact with the surface of the die, the material is locked against the tool by friction and the forming pressure, and this prevents further deformation within that region. The remaining free regions of the dome continue to deform making the formed section more uniform. Since the corners of the die are usually the last regions to be filled, the greatest strain is developed at these points. The limiting factors in simple female forming are the aspect ratio of the die (the ratio of width to depth) and the corner radii.

Simple female forming is generally preferred over

160

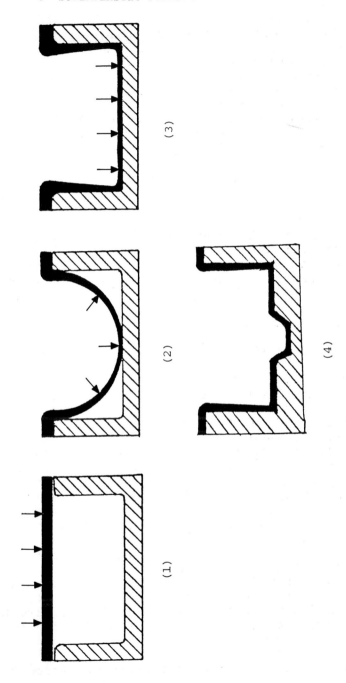

Figure 6.1 Simple female forming (schematic)

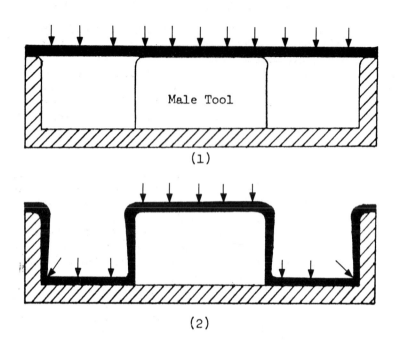

Male Tool

(1)

(2)

Figure 6.2 Female drape forming (schematic)

other forming techniques when the aspect ratio of the component is low or when the convex surface of the finished component has to be a specific shape, regardless of thickness. Stiffening features such as deep pockets or grooves can easily be incorporated into the design of parts (Fig. 6.1 (4)). Simple female forming is probably the most common method of forming.

6.2.2 Female drape forming

This process consists of bulge forming a sheet into a female mould in which one or more male details are placed (Fig. 6.2). As the sheet is bulged, the polar region will be the first to make contact with the male tool. Deformation of the superplastic sheet over the area of contact will then cease. Continued application of the forward forming pressure will drape the forming sheet over these details forcing the sheet to take on the form of the male details. The drape forming process, like simple female forming, results in a relatively uniform thickness distribution in the regions of the sheet covering the male tooling.

Female drape forming is preferred to simple female forming when dimensional tolerances of the concave surface of the finished component are important. Drape forming is also limited to parts with relatively low aspect ratios. A number of small dies can be placed within the same female forming tool allowing several parts of a component to be formed at the same time. The parts are then cut from the formed sheet and assembled into the finished component. Despite the higher wastage of material, this route can give cost savings compared with the forming of each part separately.

A number of techniques have been developed to reduce the thickness variations inherent in components formed by either of the processes described above and to enable parts with substantially higher aspect ratios than 0.3-0.4 to be fabricated.

6.2.3 Reverse bulging

In reverse bulging, the sheet to be formed is first blown away from the female mould into which it will eventually be formed. After producing the initial dome, the forming pressure is reversed forcing the bubble to fold in on itself into a female mould. During the initial bulging, deformation is concentrated primarily in the pole of the forming bubble. However, on reversing the pressure, little additional deformation occurs within that region. Instead, plastic flow is concentrated within the zone adjacent to the clamping flange and results in a much more uniform thckness distribution in the formed shape than could be obtained by simple female forming (Fig. 6.3). The height to which the

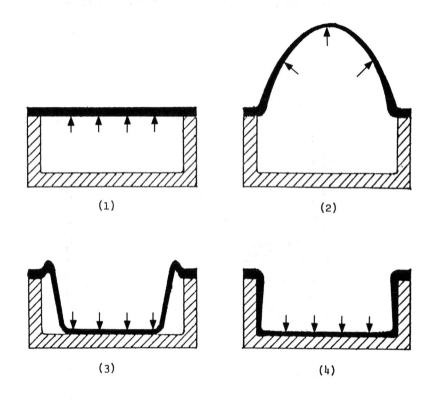

(1)

(2)

(3)

(4)

Figure 6.3 Reverse bulging (schematic)

initial bubble is blown should not be more than
about 10% greater than the depth of the final
formed component. If the height of the bubble is
too great then folds or wrinkles are likely to
develop in the finished part.

6.2.4 Plug-assisted forming

A moving auxillary tool can be used to pre-stretch
areas of the forming sheet (Fig. 6.4). As the tool
moves past the starting plane of the forming
process, contact is made between the plug and the

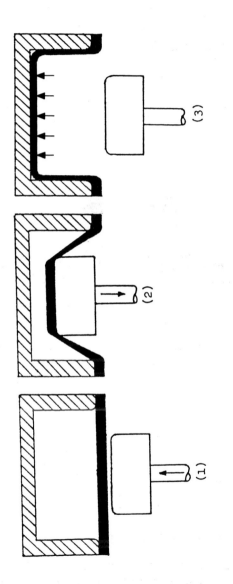

Figure 6.4 Plug-assisted forming (schematic)

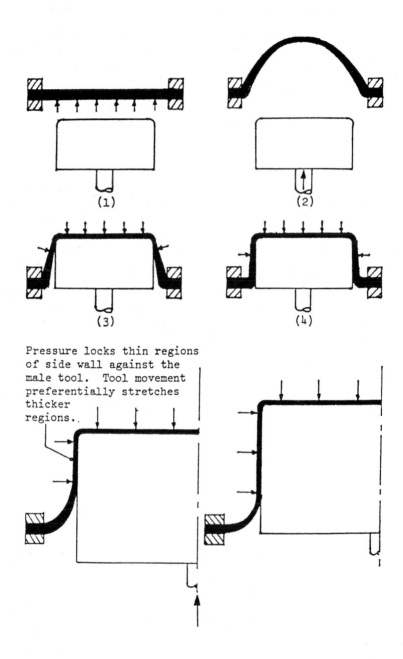

Pressure locks thin regions of side wall against the male tool. Tool movement preferentially stretches thicker regions.

Figure 6.5 Snap-back forming (schematic)

central regions of that sheet. Frictional forces prevent any significant stretching of those areas of the sheet in contact with the plug and deformation is transferred to the annular zone between the plug and the clamping flange. After pre-stretching, the plug is withdrawn and gas pressure used to complete the female forming process.

6.2.5 Snap-back forming

Snap-back forming is the male forming equivalent of plug-assisted forming, the plug defining the final form of the component (Fig. 6.5). The sheet to be formed is first blown into a bubble away from the male tool. Once the bubble has formed, the tool is moved up into the bubble. Deformation is thereby transferred from the pole to the sidewalls of the bubble. As the male plug continues to move into the bubble, the forming pressure is reversed forcing the bubble to collapse onto the plug. A combination of friction and forming pressure lock the sheet in contact with the tool and effectively prevent any further deformation. As the plug continues to move, deformation switches to the relatively undeformed material adjacent to the flange. Parts with aspect ratios of the order of 0.7, and greater, can be formed by this method.

6.3 Component Design Criteria

6.3.1 Selection of starting gauge

Component stiffness, rigidity and strength are often the foremost mechanical design criteria. Superplastic forming offers considerable scope for the incorporation of stiffening features such as ridges, grooves and bosses into components, as only one tool rather than a closely matching pair of tools are required for bulge forming. Moreover, superplastic forming can be combined with diffusion bonding to produce monolithic cellular structures with higher torsional rigidity and strength to weight ratios than could be formed by conventional fabrication methods. Considerable savings in materials, tooling and production costs are possible over the traditionally accepted routes of metal fabrication.

The attainment of specified fatigue endurance and fracture toughness levels is becoming increasingly important in materials which are to be the primary load bearing members of a structure. The strength and fatigue requirements are therefore often responsible for establishing the minimum thickness of material, h_{min}, within a component.

The starting gauge, h_o, required to meet a specified minimum thickness will depend on the component size, tool aspect ratio and the route by which it will be formed. In general

$$h_o = F_1 \frac{h_{min} \text{ x surface area of component}}{\text{plan area of component}} \qquad (6.1)$$

where F_1 is the thinning factor associated with a particular forming process (usually determined by experiment).

6.3.2 Determination of hydrostatic pressure requirements

Once the starting gauge has been established, the maximum true strain that will be attained during forming can be calculated. It is then necessary to find out whether or not the chosen material can be superplastically formed to the required strain, and to determine the appropriate minimum corner radii such that this strain is not exceeded. If the formed component is to be used in a load bearing application then the level of cavitation which will develop at the point of highest strain should be calculated, and hence the back pressure required to suppress the cavitation evaluated.

For example: It is desired to form a part in plane strain from Lital-A (Al-8090) in which the maximum true strain would be -1.3. The alloy is known to cavitate at the imposed deformation rate (8×10^{-4}/s at 520°C). Using cavitation data obtained from uniaxial tensile tests at that

168

strain rate, the volume fraction of voids, C_{vu}, at an elongation of 267% (a true equivalent plastic strain of 1.3) is found to be 0.004, for a given batch of material. Data obtained from the same set of constant strain rate tensile tests and a strain rate jump test showed that the flow stress was equal to $7.5 \, N \, mm^{-2}$ and the strain rate sensitivity, m, was equal to 0.45 at this strain rate.

The values of σ_M/σ_E for uniaxial and plane strain deformation are obtained from Table 5.2, by setting P, the superimposed pressure, equal to zero. (In this case $\sigma_M/\sigma_E = 0.5$ for uniaxial tension and 0.72 for plane strain, values half way between the limits for rigid and freely sliding grain boundaries as calculated from the models of Beere and Cocks and Ashby). The value of the cavity growth rate parameter is then calculated using equation (5.12). Hence for uniaxial deformation;

$$\eta_u = \frac{3}{2}\left[\frac{1+0.45}{0.45}\right] \sinh\left[2\left[\frac{2-0.45}{2+0.45}\right]0.50\right] = 3.27 \quad (6.2)$$

and for plane strain deformation

$$\eta_{ps} = \frac{3}{2}\left[\frac{1+0.45}{0.45}\right] \sinh\left[2\left[\frac{2-0.45}{2+0.45}\right]0.72\right] = 5.04 \quad (6.3)$$

The volume fraction of voids which would be expected to form during plane strain deformation, C_{vps}, is calculated using eqn (5.45), i.e.

$$C_{vps} = C_{vu} \exp(\epsilon(\eta_{ps}-\eta_u)) \qquad (6.4a)$$

Continuing the example and substituting $\epsilon = 1.3$ and $C_{vu} = 0.004$, the maximum volume fraction of voids that would be expected to develop during forming would be:

$$C_{vps} = 0.004 \exp(1.3(5.04-3.27)) = 0.040 \qquad (6.4b)$$

If this level of cavitation is thought to be too high (>1%) then the required back pressure is calculated from eqn (5.26), remembering that σ_E is the highest measured stress up to a strain of 1.3 for uniaxial deformation at the forming strain rate and temperature. Assuming that Lital A (Al-8090) exhibits virtually no strain hardening during superplastic flow at 520°C, the back pressure required to eliminate the cavitation damage would be (Table 5.1):

$$\frac{\sigma_M}{\sigma_E} = \left[\frac{5}{4\sqrt{3}} - \frac{P}{\sigma_E} \right] \leq 0 \tag{6.5}$$

which can be simplified to

$$P \geq 5/4\sqrt{3}\ \sigma_E \tag{6.6a}$$

i.e.

$$P \geq 0.72 \times 7.5\ \mathrm{N\,mm^{-2}}$$

$$\geq 5.40\ \mathrm{N\,mm^{-2}} \tag{6.6b}$$

If the pressure required is higher than that which can be safely applied during forming, as is likely to be the case here, then a certain amount of cavitation in the finished component must be accepted. The volume fraction of voids which would be expected to form at the maximum operating pressure can easily be calculated using equation (6.5). If for example the maximum allowable hydrostatic pressure is 3.5 MN m^{-2} (500psi) then

$$\frac{\sigma_M}{\sigma_E} = \frac{1}{2}\left[\left[\frac{1}{\sqrt{3}} - \frac{P}{\sigma_E} \right] + \left[\frac{3}{2\sqrt{3}} - \frac{P}{\sigma_E} \right]\right] \tag{6.7}$$

$$= \frac{1}{2}\left[\left[\frac{1}{\sqrt{3}} - \frac{3.5}{7.5} \right] + \left[\frac{3}{2\sqrt{3}} - \frac{3.5}{7.5} \right]\right]$$

$$= 0.255$$

Substituting the result obtained above for σ_M/σ_E into equation (5.12) and solving for $\cap$

$$\cap_{ps} = 1.59 \qquad (6.8)$$

The volume fraction of voids that would be expected at a true strain of 1.3 after forming in plane strain with a superimposed pressure of 3.5 MNm^{-2} would be

$$C_{vps} = 0.004 \exp(1.3(1.59-3.27)) = 0.0005 \qquad (6.9)$$

which is almost negligible. Alternatively, the level of cavitation which would be acceptable in the part can be set (for example $C_v = 0.005$), and the minimum hydrostatic pressure required to attain this level is evaluated for the point of maximum strain. From equation (6.4a) the required value of $\cap_{ps}$ is first calculated:

$$\cap_{ps} = \cap_u + \frac{1}{1.3}.\ln\left[\frac{0.005}{0.004}\right] = 3.27+0.17 = 3.44 \quad (6.10)$$

The next step is to calculate σ_M/σ_E from eqn (5.12).

$$3.44 = \frac{3}{2}\left[\frac{1+0.45}{0.45}\right]\sinh\left[2\left[\frac{2-0.45}{2+0.45}\right]\left[\frac{\sigma_M}{\sigma_E}\right]\right]$$

Thus
$$\sigma_M/\sigma_E = 0.523 \qquad (6.11)$$

Next, eqn (6.5) is solved for P. Thus

$$\frac{\sigma_M}{\sigma_E} = \frac{1}{2}\left[\frac{5}{2\sqrt{3}} - \frac{2P}{\sigma_E}\right] = 0.523 \qquad (6.12a)$$

i.e.

$$P = 1.49\,\text{N mm}^{-2} \quad (213\ \text{psi}) \qquad (6.12b)$$

6.3.3 Establishing the minimum corner radii

Superplastic forming, while capable of reproducing fine detail, cannot produce very sharp corners and careful consideration needs be given to determining the minimum radii of curvature that can be sustained without excessive thinning or wrinkling of the sheet. Several empirical relationships have been established to enable such radii of curvature to be determined and are given elsewhere [241-243].

6.4 Determination of the Pressure - Time Cycle for Superplastic Forming

A number of theoretical analyses have been developed to predict the pressure-time cycle required to form simple shapes and/or describe the thickness variations in the walls of those shapes [244-250].

Unfortunately, production parts frequently have complex shapes which do not readily lend themselves to even an empirical analysis of the pressure-time cycle that would be required for forming at strain rates close to the optimum for superplasticity let alone predict the resulting variations in wall thickness. However, the parts to be formed can often be approximated to one of two idealised shapes, either a circular section 'top hat' or a cylindrical/rectangular section 'trough'. The two simplified deformation geometries represent either balanced biaxial tension or plane strain forming of thin walled shells. If such simple deformation geometries are adopted then it is possible to establish an approximate pressure-time cycle from which the final forming procedure can be derived by experimentation.

6.4.1 Superplastic forming of a hemispherical diaphragm

To calculate the pressure-time cycle required to form a hemispherical shell with a final radius of r, at an approximately constant strain rate, $\dot{\epsilon}$, it will be assumed that the diaphragm maintains a

172

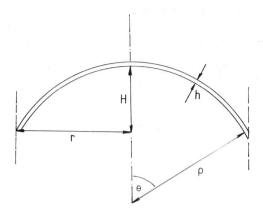

Figure 6.6 Geometry used for hemispherical bulge forming.

uniform cross-sectional thickness. The stress state within a shell of uniform thickness, h, has already been given in eqn (5.32) from which it follows that

$$\sigma_E = \frac{\delta P}{2} \left[\frac{\varrho}{h} + 1 \right] \qquad (6.13)$$

For the majority of forming operations $\varrho \gg h$; hence,

$$\sigma_E = \delta P \varrho / 2h \qquad (6.14)$$

From the geometry shown in figure 6.6 it is found that

$$\varrho^2 = r^2 + (\varrho - H)^2 \qquad (6.15)$$

$$\varrho = (r^2 + H^2)/2H \qquad (6.16)$$

The surface area of the spherical segment, A, which defines the surface of the dome is given by

$$A = 2\pi\varrho^2(1 - \cos\theta) \qquad (6.17)$$

where 2θ is the solid angle subtended by the segment at the centre of the sphere. Assuming that deformation occurs at constant volume

$$\pi r^2 h_o = 2\pi \varrho^2 (1-\cos\theta) h \qquad (6.18)$$

Recalling that $\cos\theta = (\varrho-H)/\varrho$ then

$$H = r \sqrt{(h_o/h - 1)} \qquad (6.19)$$

and

$$\varrho = \frac{r(h_o/h)}{2 \sqrt{(h_o/h - 1)}} \qquad (6.20)$$

If the dome is further assumed to maintain a uniform thickness then that thickness, h, at any time, t, is given by

$$h = h_o \exp(-\dot{\epsilon}t) \qquad (6.21)$$

From eqns (5.1) and (6.14) we obtain

$$\frac{\delta P \varrho}{2h} = \left[\frac{\dot{\epsilon}}{k'}\right]^m \qquad (6.22)$$

Substituting eqn (6.20) for ϱ and eqn (6.21) for h, the differential pressure, δP, required to form a hemispherical dome at a constant strain rate, $\dot{\epsilon}$, is obtained.

$$\delta P = 4 \left[\frac{\dot{\epsilon}}{k'}\right]^m h_o/r \exp(-2\dot{\epsilon}t) \sqrt{(\exp(\dot{\epsilon}t)-1)} \qquad (6.23)$$

If forming takes place in the presence of a hydrostatic pressure, P, then the actual forward pressure would equal $P+\delta P$. Similarly, if the material to be formed undergoes strain (grain growth) hardening during superplastic flow then the material constant k' could be replaced by an appropriate strain dependent function. The predicted pressure-time cycles for forming Al-7475 at 515°C and a strain rate of 5×10^{-4}/s are shown in figure 6.7.

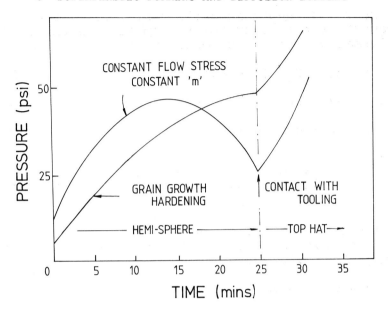

Figure 6.7 Calculated pressure-time cycle for 2mm thick Al-7475 at 515°C at a nominal strain rate of 5×10^{-4}/s, with and without the inclusion of strain hardening due to grain growth.

6.5 Diffusion Bonding

6.5.1 The DB-SPF process

Diffusion bonding is a solid state joining process in which the two surfaces to be joined are brought into contact at an elevated temperature. Application of a moderate pressure to each component brings the surfaces to be bonded into intimate contact creating a planar array of interfacial voids. Diffusion and creep flow processes can then transport atoms to the void surfaces from the adjacent areas so reducing the volume of the interfacial voids. Given sufficient

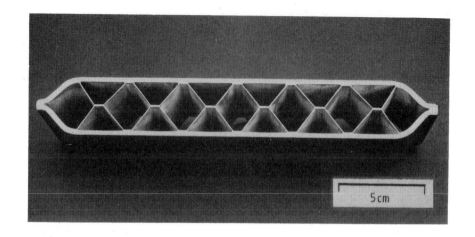

Figure 6.8 Example of cellular structures which can be formed by diffusion bonding and superplastic forming (Ti-6Al-4V).

time, the voids will disappear completely and hence an atom to atom bond is formed across the original interface. Since the bond zone is not melted during the bonding processes, the microstructure of the bond is identical to that of areas remote from the bond and has the same mechanical properties as the material itself. Furthermore, since the applied pressure required to form the bond is low there is virtually no macroscopic distortion of either component.

In the context of superplasticity, diffusion bonding is used for selective bonding of sheet material into sandwichlike constructions. Expanding the sandwich superplastically forms a shell structure. The internal structure of the shell depends on the number of interlayer sheets which make up the initial sandwich and the pattern of non-bonded or bond areas. Those shell structures which are internally stiffened mimic the cellular structures present in nature [251], and since they are virtually all empty space they have low overall densities (Fig. 6.8). The stiffened shell or cellular structures have extremely high torsional rigidity or resistance to flexure. Finally, since the structures are formed from high strength alloys, the components formed by diffusion bonding and superplastic forming have extremely high rigidity and strength-to-weight ratios.

Diffusion bonding and superplastic forming technology (DB-SPF) is readily applicable to titanium-base superplastics, since the titanium lattice has a high capacity to take into solution the surface oxide and other contaminants which would normally prevent the formation of a metal to metal bond, when the two surfaces come into intimate contact. There is considerable interest in extending DB-SPF to other superplastic alloys. However, in the case of aluminium alloys the problems associated with the tenacious surface oxide, which effectively prevents the formation of the all important metal to metal bond, has so far limited the extension of DB-SPF.

Diffusion bonded and superplastically formed structures are produced from two or more sheets of superplastic material which are selectively bonded [252-254]. The areas in which bonding is to be prevented are silk screen printed with 'stop-off' (a mixture of yttria and boron nitride in a polymeric binder). This enables the two sheets to separate at these points during the superplastic forming part of the DB-SPF process. Since the pattern is printed onto the sheets it is relatively simple to alter it, and hence effect substantial changes in the geometry of the internal stiffening elements (Fig. 6.8).

The printed sheets are assembled in register and either clamped or welded around their periphery. If the sheets are welded then prior to finally sealing the sandwich of sheets, the internal regions are evacuated. The sandwich is then heated to the diffusion bonding temperature (around 920°C, for titanium alloys such as Ti-6Al-4V), and the external surfaces subjected to a low pressure (1 to 2 MNm^{-2}). The pressure, which results in accelerated bonding, is normally applied hydraulically using compressed gas. After a predetermined period, the external pressure is released. The sandwich is then internally pressurised. This latter stage may be carried out immediately after diffusion bonding or as a separate operation. The internal pressure causes the outer skins of the sandwich to expand in opposite directions and to fill the containing cavity. By varying the shape of that cavity (i.e. making it a mould) any external shape can be produced. (The derivation of the required variation of the internal pressure with time in order that the sheets deform superplastically, was described in section 6.4). As the internal sheets are physically bonded to the external ones they are forced to deform as the structure expands and the internal stiffening structure is thus developed.

6.5.2 Bonding mechanisms

Since no surface is atomically smooth the placing of two surfaces together will result in a finite number of contact points and the creation of a planar array of highly irregular voids. If the points of contact are atomically clean then a metal bond will form. The bulk of the time required for the diffusion bonding process is spent in bringing the remaining surfaces, which are not within an atomic diameter of each other, into close proximity. In the same manner as the application of pressure speeds up conventional sintering, the application of pressure enhances the rate at which the unbonded surfaces approach. In this context the bonding process is that which brings the two surfaces into atomic contact, rather than the process by which the interatomic bond itself forms. Thus the time for bonding is that required to attain 100% contact between the two surfaces, the actual bond being assumed to form instantaneously on such contact.

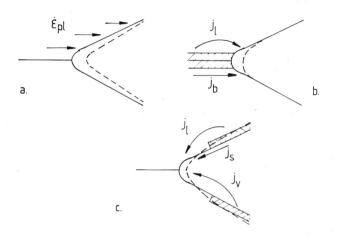

Figure 6.9 Schematic illustration of the mass
transfer paths which can operate in order to fill
the interfacial voids during diffusion bonding.
(a) time dependent plastic collapse of the
interface, (b) diffusion from interfacial sources
to the void surface and (c) surface diffusion.

The mechanisms by which the interfacial voids are
removed are as follows (Fig. 6.9):

1. Instantaneous plastic collapse. When the
   bonding pressure, P, is applied the points of
   contact between the two surfaces will be unable to
   support the applied stress. The contact area will
   therefore expand almost instantaneously until the
   stress acting perpendicular to the contact area is
   less than the yield stress, σ_y. This is taken as
   the starting point for bonding. At this instant,
   the area fraction bonded, f_a, is given by

   $$f_a = P/\sigma_y \qquad (6.24)$$

2. Time dependent plastic collapse. The stress
   within the contact zone will cause plastic flow by

either conventional creep or superplasticity. Plastic flow will force material into the voids and thereby expand the contact area (Fig. 6.9a). The stress in the bonded region will decrease as the area fraction bonded increases, and hence the rate of void closure by time dependent plastic collapse will lessen with increasing time.

3. Diffusion. During application of the bonding pressure, the contact interface between the two sheets experiences a normal stress. Since the free surface of the void cannot support a normal stress there exists a difference in chemical potential between the interface and the void surface. The gradient in chemical potential drives diffusion such that atoms move from the contact interface to the void surface and the void volume decreases accordingly. The mass transfer path can be both via the interface itself (boundary diffusion) or through the lattice (volume diffusion) (Fig. 6.9b) or via the surface or vapour phase (Fig. 6.9c).

6.5.3 Kinetics of diffusion bonding

The first attempt to predict the time required to form a sound bond between two rough surfaces was made by Hamilton [255]. It was assumed that the surfaces to be bonded consisted of triangular section asperities in point-to-point contact (Fig. 6.10). Hamilton calculated the average strain rate in the section, based on the average net section stress through eqn (4.1). The bonding time, t, was calculated by dividing the strain required to bring the asperities into an overlapping position $(\epsilon_z = -0.693)$ by the average strain rate. Since the stress is proportional to the applied pressure, P, then

$$ t = \frac{0.8kT}{A_{II} \ D_{gb}\delta G} \left[\frac{G}{\sqrt{3}P} \right]^n \tag{6.25} $$

This simple model was later extended to describe a more realistic surface geometry. In the later model [256] the surfaces to be bonded were considered to

180

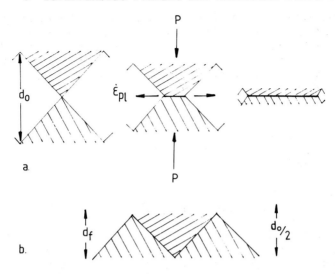

Figure 6.10 Geometry adopted by Hamilton [255] to model diffusion bonding in plane strain.

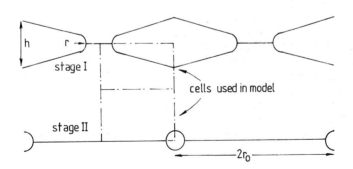

Figure 6.11 Geometry adopted by Derby and Wallach [257] to model diffusion bonding in plane strain.

consist of short wavelength asperities superimposed on a longer wavelength roughness. The bonding process was divided into two stages. Firstly, a reduction in height of the long wavelength roughness, as in the original model of Hamilton, was assumed to take place until the height of the remaining voids was equal to that of the short wavelength asperity amplitude. This occurred by time dependent plastic collapse only. In the second stage of the bonding process, interfacial diffusion was believed to be the dominant void closure mechanism and the time to sinter the remaining planar array of microvoids was calculated. The total bonding time was taken to be the sum of the two sequential processes.

A further refinement of the plane strain model of diffusion bonding was introduced by Derby and Wallach [257,258], who considered diffusion bonding to be analogous to pressure sintering [259]. Again the bonding process was divided into two stages, I, the collapse of a ridge and, II, the sintering of cylindrically symmetric voids (Fig. 6.11). Unlike the previous model, all the bonding mechanisms were considered to operate simultaneously. During the initial stages of bonding, the curvature of the neck of the void was much greater than that of the surfaces remote from the neck. Again a chemical potential gradient resulted in surface diffusion of mass to the neck from regions with lower surface curvature. Hence the void radius in the plane of the bond interface would decrease due to surface diffusion. However, since surface diffusion only redistributes matter within the void, no change in void volume would occur. If surface diffusion is very rapid then the voids would always maintain a cylindrically symmetrical geometry. The rate of change of void radius with time is then given by

$$\frac{dr}{dt} = -\sqrt{3}\pi r_o^2 (1-f_a)^2 A_{II} \frac{D_{gb}\delta G}{kT} \left[\frac{2P}{2f_a (1-(1-f_a)^{2/n})} \right]^n$$

$$- \frac{3\Omega P}{f_a^2 r_o kT} \left[\frac{D_{gb}\delta}{2} + D_v r_o (1-f_a) \right] \qquad (6.26)$$

182

in which f_a is the area fraction of the
interface already in contact (bonded), P the
applied pressure and r_o the original surface
roughness.

For superplastic materials, the grain size can be
substantially smaller than the surface roughness
which results from cleaning processes such as
brushing or grit blasting. In the model of Derby
and Wallach it was assumed that only one
interface, the bond interface, intersected the
surface of the voids and hence only one plane of
diffusive mass transfer was available. However,
when the grain size is smaller than the void size,
numerous grain boundary interfaces will intersect
the void surface and a number of additional mass
transfer paths for diffusion become available [211].
In effect this is the reverse of superplastic
diffusional void growth. For cylindrically
symmetrical voids the rate of change of the void
radius with time is given by a modification of
eqn (6.26) in which the second term, representing
the diffusive flux along the bond interface, is
replaced by a summation of diffusive fluxes arising
from all the interfaces intersecting the void.
Naturally, as the void becomes smaller then the
number of grain boundary interfaces which intersect
the void surface will be reduced. The number of
additional interfaces, N, which must be an
integer, is given by

$$N = 2 \ \text{int}(r_o(1-f_a)/d) \qquad (6.27)$$

where d is the grain size. In making the
summation, the area fraction bonded, f_a, must be
replaced by the apparent area fraction bonded in
the plane parallel to the bond interface, where the
the grain boundary interface intersects the void
surface (Fig. 6.12). The time required to attain
full interfacial contact, t, is then

$$t = -r_o \int_{P/\sigma_y}^{1} \frac{df_a}{\Sigma \ dr/dt} \qquad (6.28)$$

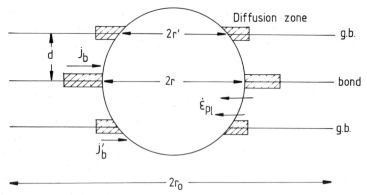

Figure 6.12 Geometry adopted by Pilling et al. to model diffusion bonding in plane strain [211].

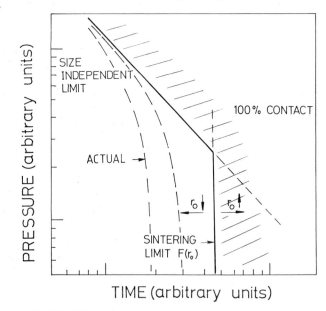

Figure 6.13 The calculated variation of bonding time with pressure at a fixed temperature. At low pressure, the solution is independent of pressure, but strongly dependent on the surface finish (r_o). At high pressure, the bonding time depends exponentially on the pressure and is independent of the surface finish. Combinations of pressure and time to the right of the solid line (shaded area) will result in complete contact between the two surfaces being joined.

where dr/dt is expressed as function of f_a, the material constants (as per equation (6.26)), the applied pressure, P, temperature, T, and the initial surface roughness, r_o.

In some DB manufacturing processes, the sheets to be bonded are welded around their peripheries and placed in a sealed box which can be pressurised. The bonding process is no longer one of plane strain, but is isostatic. While the same physical mechanisms of void closure are operative, the kinetics of each process will differ from that under plane strain conditions. On average, the rate of void closure during isostatic bonding is a factor of 2 to 3 times greater than that for plane strain bonding [260]

In deriving the rate equations for each mechanism and model it is important to define the geometry of the interfacial voids and the applied stress state in such a way that there is no discontinuity in the rate equations when the voids become isolated from each other. Moreover, it is important to recognise that the surfaces being joined have a complex topography. Often a short wavelength, low amplitude roughness is superimposed on a much longer wavelength, low aspect ratio, waviness and it is normally the latter aspect of the surface geometry that determines the overall bonding time,(Fig. 6.13).

Given the uncertainties in the material property data (a factor of 2 to 5 for most values), the times for bonding that are predicted by the models should not be regarded as absolute, but rather as a guide to the 'process window' in which experimentation should be carried out. By isolating the time frame in which bonding would be expected to occur under a given set of bonding conditions (pressure, temperature, surface finish and material) actual experimentation and destructive evaluation of the bond properties can be minimised.

6.5.4. Testing of Diffusion Bonds

The quality of diffusion bonds can only be reliably assessed by comparing the fracture characteristics of the bond with those of the parent metal after a simulated bonding heat treatment cycle. Ultrasonic evaluation of the bond quality is presently incapable of resolving bond line microvoids (d<5 μm) since the wavelength of the sound wave (~200 μm) is much greater than the defect size and, in thin sheets, the transit times are extremely short. More recently, improvements in signal processing have enabled some headway to be made in detecting disbonds in which the two surfaces are often in intimate contact [264]. Once the conditions required to produce an acceptable bond quality have been determined by destructive testing, strict process control is normally employed to ensure reproducibility of the bond properties. This is often supplemented by proof testing of lugs which have been designed on to the component and which are removed after the bonding cycle is complete [265].

A number of mechanical tests have been devised to assess the quality of diffusion bonds. In the case of bonds produced in thick sections, conventional tensile, rotating bend fatigue and impact test pieces can be readily produced and the fracture strengths compared with that of the parent material. In practice, parent metal tensile strengths are often achieved in bonds with more than 85% interfacial contact, while parent metal fatigue endurance normally requires a complete absence of interfacial microvoids. However, the most discriminating measure of bond quality is obtained by impact testing, when poor bonds which show complete interfacial contact may exhibit low impact strengths.

Unfortunately, impact testing is not widely applicable because the majority of diffusion bonds are formed between sheet materials. In such cases, the fracture strength of the joint is measured using a constrained lap shear test although no test piece standard exists. Furthermore, the attainment of parent metal fracture strength is no guarantee that

186

a high quality bond has been formed, nor that the bond will be capable of resisting peel during superplastic forming. It is this latter capability which might prove to be the most discriminating test for evaluating diffusion bonds between thin sheets.

6.5.5 Diffusion Bonds in Titanium Alloys

Titanium alloys such as Ti-6Al-4V, Ti-6Al-2Sn-4Zr-2Mo and Ti-4Al-4Mo-2Sn-0.5Si (IMI 550) are readily bonded at temperatures between 880 and 940°C with applied pressures of 0.6 to 2MPa for process times up to 3 hours. The temperatures and stresses used for bonding are normally close to the conditions under which optimum superplasticity is observed, Table 2.3. In general, as the bonding temperatures and times are increased, the integrity of the bonds increases. However, excessive temperatures and long process times can lead to grain growth during the bonding cycle and a reduction in the subsequent superplastic formability of the alloys. The relative ease with which solid state diffusion bonds can be formed in titanium-base alloys is reflected in the growing body of literature and its acceptance as a mainstream manufacturing technology [252,253,266-269].

The ease with which titanium can be bonded may be attributed to the ability of titanium to take into solution both its oxide and other surface contaminants when subjected to pressure at elevated temperatures [270]. It has been found that parent metal fatigue strengths can be attained in titanium diffusion bonds [268], figure 6.14, but that impact strengths are often less than those of the parent metal [271], (Fig. 6.15). The latter can be improved by heating to the β transus after bonding is complete, allowing grain boundary migration, and hence a disruption of the planar bond interface, to occur.

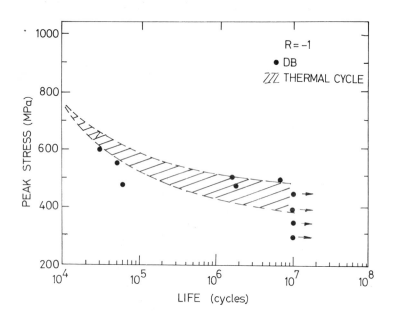

Figure 6.14 Fatigue properties of Ti-6Al-4V after bonding for 2 hours at 950°C under an applied pressure of 0.69 MPa (100psi) [268].

6.5.6 Diffusion Bonding of Aluminium Alloys

Unlike titanium alloys, the development of a DB/SPF technology for aluminium alloys has proved difficult because of the existence of a tenacious surface oxide However, studies have shown that 'solid state' diffusion bonds with fracture strengths in excess of those attained using polymeric adhesives can be realised. Two types of bonding procedure have emerged. Firstly, that involving surface modification, such as the deposition of a silver coating on a nominally oxide-free aluminium surface and secondly, the use of large scale deformation in the bond zone to fracture the tenacious layer of alumina on the aluminium surface.

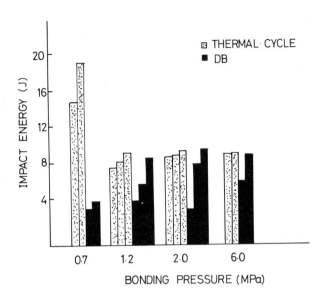

Figure 6.15 Impact strength of IMI-550 (see Table 2.3) after bonding for 2 hours at 950°C under various applied pressures [271].

Silver coated surfaces have been produced on clad Al-7010 [272,273], Al-2014 and Al-6082 [274], Al-5056, Al-5657 and Al-6351 [275]. The coating prevents the formation of alumina and any surface oxidation will lead to the formation of a silver oxide (Ag_2O) which is soluble under pressure at elevated temperatures. The aluminium surfaces to be coated are polished to <1 µm finish then sputter cleaned in vacuum using argon ions. The silver coating is applied either by vapour deposition or by silver ion plating.

Bonding is generally carried out in vacuum at pressures ranging from 1.5 to 5 MPa at temperatures

of 450 to 550°C for times up to 1 hour. For the clad
Al-7010, bonding has been carried out at 200 to
300°C using pressures of 130 to 140 MPa. In all
cases, the bond shear fracture strength increased
with an increase in the bonding temperature and the
extent of deformation in the bond zone. Despite the
large number of alloys and conditions investigated,
bond shear fracture strengths in the range 50 to 80
Nmm^{-2} were generally observed. Post-bonding heat
treatment, usually to the T6 or equivalent
condition, resulted in increased bond fracture
strengths of 140 to 180 N mm^{-2}.

Solid state bonding without the use of a protective
coating has been restricted to two alloys, Supral
220 (a fine grain derivative of Al-2004) [276] and
Al-7475 [277-279]. Bonding temperatures and times
were limited by the amount of grain growth which
could be tolerated during the bonding cycle before
the subsequent SPF behaviour of the material was
impaired. Bonds which were fabricated under static
compressive loading gave shear fracture strengths in
the range 40 to 150 Nmm^{-2}. Again, the bond
strength could be improved by the application of a
post-bonding heat treatment with shear fracture
strengths as high as 250 Nmm^{-2} being recorded.

The conditions under which solid state bonds have
been produced in aluminium alloys are summarised in
Table 6.1 together with the reported fracture
strengths.

A frequent and disquieting feature of solid state
bonds formed between two aluminium surfaces has been
the wide variation in the measured fracture strength
of bonds formed under nominally the same conditions.
For example, Byun and Yavari [277], who examined 3
to 4 bonds per set of process variables recorded
deviations of +/- 15% while Pilling and Ridley
[279], who examined between 16 and 20 bonds per set
of process variables, observed deviations of +100%
to -60% of the mean fracture strength. In the latter
study, the variability of the measured bond fracture
strength was analysed using Weibull statistics;
Weibull modulii of 4 to 6 were typical of the
inherent variability of the bond fracture strength

Table 6.1 Summary of the conditions under which diffusion bonds have been produced between aluminium alloys together with the fracture strengths of the bonds formed. (τ, shear; σ, tensile).

Material	Surface Finish	Coating	Temp °C	Pressure MPa	Time h	Strength, MPa	Ref.
clad 7010	polished	Ag ion-plated	220-300	130-140	½	74-80 (τ)	272,273
Al	polished	Ag ion-plated	500-550	1.5	<1	53-74 (τ)	274
5056	polished	Ag vapour dep	450-550	2.5-5.0	<1	50-145 (σ)	275
7475	grit blasted	none	515-550	2-20	2-12	40-150 (τ)	277,278
7475	ground	none	510	2.76	1	50-150 (τ)	
2004	grit blasted	none	460-490	1-5	4-6	55-105 (τ)	276
7475	grit blasted	Zn vapour dep	515	5	5	40-110 (τ)	276
7475	ground	Al-12Si foil	510	1.38	½	150 (τ)	278

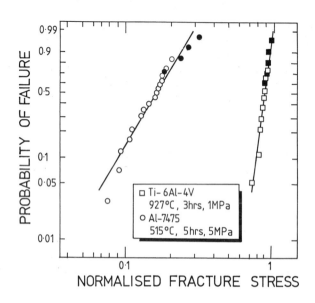

Figure 6.16 Probability of fracture, as determined by Weibull statistics versus normalised fracture strength (τ_F/τ_{UTS}) for Al-7475 and Ti-6Al-4V.

(Fig. 6.16). (By way of comparison, repeated measurements of the yield strength of copper would give a Weibull modulus of ≈40, while that of the modulus of rupture of a structural ceramic would be closer to 10).

Attempts have been made to bond aluminium using a transient liquid phase, although this is not strictly diffusion bonding. By interposing a melting point depressant between the two aluminium surfaces it is possible to form a liquid layer in the bond zone during the bonding cycle. The liquid phase forms directly by melting on heating or by isothermal melting at the bonding temperature as the concentration of the melting point depressant builds

192

up in the surface layer of the aluminium alloy. Application of a moderate pressure after the liquid phase has formed displaces the bulk of the liquid from the bond zone and disrupts the oxide. During the remainder of the bonding cycle the concentration of the melting point depressant decreases as it diffuses further away from the bond zone which isothermally solidifies.

Only elements which are compatible with the base alloy are suitable for forming a transient liquid phase. In the case of aluminium the choice is limited to silver, magnesium, silicon and zinc (copper, although compatible, is likely to cause intergranular embrittlement of the aluminium alloys). The melting point depressants are usually interposed between the aluminium surfaces as elemental coatings [275,277] or as aluminium alloy foils [277].

6.5.7 Diffusion Bonding of Nickel Alloys

After titanium and aluminium, nickel-base alloys are, perhaps, the next most important group of superplastic materials, see Table 2.5. However, their use has been concentrated in 'massive' rather than thin sheet components, where they have been superplastically forged into complex shapes [280,281]. Diffusion bonding of nickel (super)alloys has recently received considerable attention particularly as a means of joining superplastic alloys of dissimilar compositions, mechanically alloyed materials, and single crystal or directionally solidified materials.

Nickel, like titanium, is capable of taking into solution its surface oxide when under pressure at elevated temperatures, although other surface contaminants are not as readily soluble. Nickel-base superalloys are normally bonded at temperatures just below the τ' solvus. Depending on the alloy, bonding can be carried out at temperatures between 1100°C and 1250°C with applied pressures in the range 1 to 20 MPa. These conditions correspond to those where superplasticity can be exhibited. The mechanical strength of bonds formed between dissimilar materials can be degraded by a number of factors which would not normally be apparent when bonds are formed between like alloys. These are [282]:-

1. Penetration of the single phase τ microstructure into the two phase τ-τ' microstructure as a result of the diffusion of τ' forming elements into the single phase alloy.
2. Formation of TiC particles/films at the interface as titanium reacts with adsorbed CO on the mating surfaces.
3. Development of Kirkendall porosity on the aluminium-rich side of the joint.

The latter factor would not normally be apparent after fabrication as the isostatic pressure applied to form the bond is often more than sufficient to prevent the formation of Kirkendall porosity due to the flux imbalance between Ni and Al [283]. However, during subsequent service at elevated temperatures porosity will develop rapidly in the absence of a confining pressure [283]. The formation of carbides within the bond zone can be avoided by treating the mating surfaces with NH_4BF_4 which decomposes on heating, reacting with TiC to form volatile titanium fluoride and fluorocarbon [284].

It has been demonstrated that nickel-base alloys such as IN-718 can be diffusion bonded and superplastically formed [285] in a similar manner to superplastic titanium alloys.

6.6. Summary

Diffusion bonding and superplastic forming is now an established manufacturing technology. Its application has enabled low weight-high stiffness cellular structures to be fabricated simply and cheaply using titanium-base alloys. Although it is possible to join superplastic aluminium alloys using both solid state and transient liquid phase techniques, the poor reproducibility of each of those processes has so far limited the introduction of DB-SPF of aluminium alloys. Both superplastic and non-superplastic nickel-base alloys can be successfully diffusion bonded. However, where dissimilar materials are to be used at elevated temperatures, careful consideration needs to be given to composition to ensure that Kirkendall porosity does not become evident during service.

Acknowledgements

The authors are grateful to Dr. D.J.Lloyd and Professor T.G.Langdon for supplying several of the micrographs used in the text.

References

1   M.M.I.Ahmed and T.G.Langdon, Metall.Trans. 8A (1977) 1832-1833.
2   K.Higashi, T.Ohnishi and Y.Nakatami, Scripta Metall. 19 (1985) 821-824.
3   F.Hargreaves, J.Inst.Metals 39 (1928) 301-327.
4   F.Hargreaves and R.J.Hills J.Inst.Metals 41 (1929) 257-283.
5   C.M.H.Jenkins, J.Inst.Metals 40 (1928) 41-54.
6   C.E.Pearson, J.Inst.Metals 54 (1934) 111-124 .
7   A.A.Bochvar and Z.A.Sviderskaya, Izvest.Akad.Nauk. 9 (1945) 821.
8   E.E.Underwood, J.Metals 14 (1962) 914-919.
9   W.A.Backofen, I.R.Turner and D.H.Avery, Trans. ASM 57 (1964) 980-990.
10   K.A.Padmanabhan and G.J.Davies, Superplasticity publ. Springer-Verlag, Berlin (1980).
11   J.Hedworth and M.J.Stowell, J.Materl.Sci. 6 (1971) 1061-1069.
12   T.G.Langdon, Metall.Trans. 13A (1982) 689-701.
13   J.G.Wang and R.Raj, J.Amer.Ceram.Soc. 67 (1984) 385-390 and 399-409.
14   F.Wakai and H.Kato, Adv.Ceram.Materls. 3 (1988) 71-76.
15   T.E.Chung and T.J.Davies, Acta Metall. 27 (1979) 627-635.
16   D.Lee, Acta Metall 17 (1969) 1057-1069.
17   M.Suery and B.Baudelet, Philos.Mag. 41A (1980) 41-64.
18   A.K.Ghosh and C.H.Hamilton, Metall.Trans. 10A (1979) 699-706.
19   B.M.Watts and M.J.Stowell, J.Materl.Sci. 6 (1971) 228-237.
20   O.N.Senkov and M.M.Myshlyaev, Acta Metall. 34 (1986) 97-106.
21   M.Suery and B.Baudelet, J.Materl.Sci. 8 (1973) 363-369.
22   M.A.Clark and T.H.Alden, Acta Metall. 23 (1975) 1195-1206.

23   C.H.Cacares and D.S.Wilkinson, Scripta Metall. 16 (1982) 1363-1365.

24   G.R.Yoder and V.Weiss, Metall.Trans. 3A (1972) 675-681.

25   C.H.Caceres and D.S.Wilkinson, J.Materl.Sci. Letters 3 (1984) 395-399.

26   B.Walser and O.D.Sherby, Metall.Trans. 10A (1979) 1461-1471.

27   T.Oyama, J.Wadsworth, M.Korchynsky and O.D.Sherby, ICSMA 5, Ed. P.Hassen, V.Gerold and G.Kostoroz, publ. Pergamon Press, London (1979) 381-386.

28   J.Wadsworth and O.D.Sherby, J.Metal Working Technol. 2 (1978) 53.

29   D.Lee and W.A.Backofen, Trans. TMS-AIME 239 (1966) 1034-1040.

30   N.E.Paton and C.H.Hamilton, Metall.Trans. 10A (1979) 241-257.

31   T.Oyama, O.D.Sherby, J.Wadsworth and B.Walser, Scripta Metall. 18 (1984) 799-804.

32   J.Wadsworth and O.D.Sherby, J.Materl.Sci. 13 (1978) 2645-2649.

33   O.D.Sherby, B.Walser, C.M.Young and E.M.Cady, Scripta Metall. 9 (1975) 569-574.

34   G.Piatti, G.Pellergrini and R.Trippodo, J.Materl.Sci. 11 (1976) 186-189.

35   D.M.Moore and L.R.Morris, Materl.Sci. and Eng. 43 (1980) 85-92.

36   F.W.Ling and D.Laughlin, Metall.Trans. 10A (1979) 921-928.

37   T.H.Alden and H.W.Schadler, Trans. TMS-AIME 242 (1968) 825-832.

38   A.Ball and M.M.Hutchison, Metal Sci.J. 3 (1969) 1-7.

39   B.M.Watts, M.J.Stowell, B.L.Bakie and D.R.E.Owen, Metal Sci. 10 (1976) 189-206.

40   D.J.Lloyd and D.M.Moore, Superplastic Forming of Structural Alloys, Ed. N.E.Paton and C.H.Hamilton, publ. TMS-AIME, Warrendale (1982) 147-172.

41   T.Sakai and J.J.Jonas, Acta Metall. 32 (1984) 189-209.

42   B.Geary, J.Pilling and N.Ridley, Superplasticity in Aerospace - Aluminium Ed. R.Pearce and L.Kelly, publ. School of Industrial Science, Cranfield (1985) 127-135.

43   D.L.Holt and W.A.Backofen, Trans. ASM 59 (1966) 755-768.

44   K.A.Padmanabhan and G.J.Davies, Metal Sci. 11 (1977) 177-184.

45   R.H.Bricknell and A.P.Bentley, J.Materl.Sci. 14 (1979) 2547-2554.

46   G.Rai and N.T.Grant, Metall.Trans 6A (1975) 385-390.

47   R.Grimes, C.Baker, M.J.Stowell and B.M.Watts, Aluminium 51 (1975) 11.

48   B.Geary, Ph.D. Thesis, University of Manchester (1985).

49   K.Ohori, H.Watanabe, M.Mukaiya and Y.Endo, Al-aro 7 (1984) 17.

50   K.Matsuki, Y.Uetani M.Yameda and Y.Murakami, Metal Sci. 10 (1976) 235-242.

51   R.H.Bricknell and J.W.Edington, Metall. Trans 7A (1976) 153-155.

52   K.Matsuki, H.Morita, M.Yameda and Y.Murakami,
Metal Sci. 11 (1977) 156-163.

53   C.H.Hamilton, C.C.Bampton and N.E.Paton,
Superplastic Forming of Structural Alloys,
Ed. N.E.Paton and C.H.Hamilton,
publ. TMS-AIME, Warrendale (1982) 173-189.

54   C.C.Bampton, J.A.Wert and M.W.Mahoney, Metall.Trans
13A (1982) 193-198.

55   P.G.Partridge and A.J.Shakeshaff, Technical Rep. 82117,
Royal Aircraft Establishment, Farnborough (1982).

56   N.Ridley, B.Geary and J.Pilling, Unpublished Research,
University of Manchester, 1986.

57   J.Wadsworth, I.Palmer, D.D.Crooks and R.E.Lewis,
2nd. Int. Aluminium-Lithium Conf. Ed. E.S.Starke and
T.H.Saunders, publ. TMS-AIME, Warrendale (1983) 111-135.

58   J.Wadsworth, Metall.Trans. 16A (1985) 2312-2332.

59   J.A.Wert, N.E.Paton, C.H.Hamilton and M.W.Mahoney,
Metall.Trans. 12A (1981) 1267-1276.

60   R.Grimes, Superplasticity Ed. B.Baudelet and
M.Suery, publ. CNRS Paris (1985) Ch.13.

61   D.W.Chung and J.R.Cahoon, Metal Sci. 13 (1979)
635-640.

62   M.Otsuka, Y.Miura and R.Horiuchi, Scripta Metall.
8 (1974) 1405-1408.

63   N.Ridley, Superplastic Forming of Structural Alloys,
Ed. N.E.Paton and C.H.Hamilton, publ. TMS-AIME,
Warrendale (1982) 191-207.

64   B.Walser and U.Ritter, Superplasticity Ed. B.Baudelet
and M.Suery, publ. CNRS, Paris (1985) Ch.15.

65   J.Wadsworth, J.H.Lin and O.D.Sherby, Metal Technol.
8 (1981) 190-193.

66   C.I.Smith, B.Norgate and N.Ridley, Metal Sci. 10
(1976) 182-188.

67   N.Ridley and L.B.Duffy, ICSMA 7, Ed. H.J.McQueen et al.
publ. Pergamon Press, Oxford (1985) 853-858.

68   E.U.Engstrom, Duplex Stainless Steels '86',
Ed. J.Van Liere, publ. Nederlands Instituut voor
Lastechniek,The Hague (1986) 303-307.

69   Y.Maehara and Y.Ohmori, Metall.Trans 18A (1987) 663-672.

70   A.R.Marder, Trans. TMS-AIME 245 (1969) 1337-1342.

71   C.I.Smith and N.Ridley, Metal Technol. 1 (1974)
191-198.

72   S.K.Srivastava, Duplex Stainless Steels, publ.
ASM, Metals Park, USA (1982) 1-14.

73   W.Schadler, Trans. TMS-AIME 242 (1968) 1281-1287.

74   A.Arieli and A.Rosen, Metall.Trans. 8A (1977) 1591-1596.

75   O.A.Kaibyshev, I.V.Kazachkov and R.M.Galeev,
J.Materl.Sci. 16 (1981) 2501-2506.

76   M.T.Cope, D.R.Evetts and N.Ridley, J.Materl.Sci. 21
(1986) 4003-4008.

77   C.H.Hamilton, Superplastic Forming Ed. S.P.Agrawal,
publ. ASM, Metals Park (1985) 13-22.

78   J.Wert and N.E.Paton, Metall.Trans. 14A (1983) 2535-2544.

79   A.K.Ghosh and C.H.Hamilton, Metall.Trans. 10A
(1979) 241-250.

80   M.T.Cope and N.Ridley, Materl.Sci and Technol. 2
(1986) 140-145.

81   J.R.Leader, D.F.Neal and C.Hammond, Metall.Trans.
     17A (1986) 93-106.
82   J.Ma, R.Kent and C.Hammond, J.Materl.Sci. 21 (1986)
     475-487.
83   D.S.McDarmaid, Mat.Sci and Eng. 70 (1985) 123-129.
84   P.Griffiths and C.H.Hammond, Acta Metall. 20 (1972)
     935-945.
85   F.Wakai, S.Sakaguchi and Y.Matsuno, Adv.Ceram. Materl.
     1 (1986) 259-263.
86   F.Wakai, H.Kato, S.Sakaguchi and N.Murayama,
     Yogyo-Kykai-Shi 94 (1986) 1017-1020.
87   J.H.Hensler and G.V.Cullen, J.Amer.Ceram.Soc. 51
     (1967) 584-585.
88   J.C.Crampon and B.Escaig, J.Amer.Ceram.Soc. 63
     (1980) 680-686.
89   K.R.Venkatachari and R.Raj, J.Amer.Ceram.Soc 69
     (1986) 135-138.
90   R.M.Canon, W.H.Rhodes and A.H.Heuer, J.Amer.Ceram.Soc
     63 (1980) 46-63.
91   P.C.Panda, R.Raj and P.E.D.Morgan, J.Amer.Ceram.Soc.
     68 (1985) 522-529.
92   M.F.Merrick, Superplastic Forming of Structural
     Alloys, Ed. N.E.Paton and C.H.Hamilton,
     publ. TMS-AIME, Warrendale (1982) 209-223.
93   R.C.Gibson and J.H.Brophy, Ultrafine Grain Metals,
     Ed. J.J.Burke and V.Weiss,Publ. Syracuse Univ. Press
     (1969) 377-394.
94   E.E.Brown, R,C.Boettner and D.L.Ruckie, Superalloys -
     Processing - 2nd. Int.Conf on Superalloys,
     Publ. AIME (1972) L1-L12.
95   H.W.Hayden, R.C.Gibson, H.F.Merrick and J.H.Brophy,
     Trans. ASM. 60 (1967) 3-13.
96   B.H.Kear, J.M.Oblak and W.A.Owcarski, J.Metals 246
     (1972) 25-32.
97   D.A.Woodford, Metall.Trans. 7A (1976) 1244-1245.
98   R.L.Athey and J.B.Moore, Powder Metallurgy for High
     Performance Applications Ed. J.J.Burke and V.Weiss,
     publ. Syracuse Univ. Press (1972) 281-293.
99   S.H.Reichman and J.W.Smythe, Int.J.Powder.Met.
     6 (1970) 65-74.
100 R.G.Menzies, J.W.Edington and G.J.Davies, Metal Sci.
     15 (1981) 210-216.
101 J.P.Immarigeon and P.H.Floyd, Metall Trans. 12A
     (1981) 1177-1186.
102 J.K.Gregory, J.C.Gibeling and W.D.Nix, Superplastic
     Forming of Structural Alloys, Ed. N.E.Paton and
     C.H.Hamilton, publ TMS-AIME, Warrendale (1982) 361.
103 A.Y.Kandeil, J.P.Immarigeon, W.Wallace, Met.Sci.
     14 (1980) 493-499.
104 Y.G.Nakagawa, H.Yoshizawa and H.Terashima,
     Materl.Sci. and Technol. 2 (1986) 637-639.
105 D.W.Livesey and N.Ridley, Metall.Trans. 9A (1978)
     519-526.
106 D.W.Livesey and N.Ridley, Metall.Trans 13A (1982)
     1619-1626.
107 R.D.Schelling and G.H.Reynolds, Metall Trans 4A
     (1973) 2199-2203.

108   D.M.Ward, B.J.Helliwell and R.J.Penrice, Metallurgica
      and Metal Forming (1973) 319-324.
109   S.A.Shei and T.G.Langdon, Acta Metall. 26 (1978)
      639-646.
110   S.A.Shei and T.G.Langdon, Acta Metall. 26 (1978)
      1153-1158.
111   S.A.Shei and T.G.Langdon, J.Materl.Sci. 16 (1981)
      2988-2996.
112   R.G.Fleck, C.J.Beevers and D.M.R.Taplin, J.Materl.Sci.
      9 (1974) 1737-1744.
113   S.W.Zehr and W.Backofen, Trans.ASM. 61 (1968) 300-313.
114   B.Baudelet and M.Suery, J.Materl.Sci. 7 (1972) 512-516.
115   R.C.Gifkins,J.Inst.Met (1967) 373.
116   F.A.Mohamed and T.G.Langdon, Acta Metall. 29 (1981)
      911-920.
117   H.Naziri and R.Pearce, Int.J.Mech.Sci. 12 (1970)
      513-521.
118   H.E.Cline and D.Lee, Acta Met. 18 (1970) 315-323.
119   A.K.Ghosh, Superplastic Forming of Structural
      Alloys, Ed. N.E.Paton and C.H.Hamilton, publ.
      TMS-AIME, Warrendale (1982) 85-103.
120   S.H.Vale, D.J.Eastgate and P.M.Hazzledine, Scripta
      Metall. 13 (1979) 1157-1162.
121   R.Lagneborg, B.H.Forsen and J.Wiberg, Creep
      Strength in Steels and High Temperature Alloys,
      publ. Metal Society, London (1974) 1-7.
122   F.W.Crossman and M.F.Ashby, Acta Metall. 23 (1975)
      425-440.
123   H.J.Frost and M.F.Ashby, Deformation Mechanism Maps,
      publ. Pergamon Press Oxford (1982).
124   R.L.Coble, J.Appl.Phys. 34 (1963) 1679-1682.
125   C.Herring, J.Appl.Phys. 21 (1950) 437-445.
126   P.M.Hazzledine and D.E.Newbury, Grain Boundary
      Structure and Properties, Ed. G.A.Chadwick and
      D.A.Smith, Publ. Academic Press (1976) 235-266.
127   A.K.Mukherjee, Materl.Sci and Eng. 8 (1971) 83-89.
128   A.K.Mukherjee, Ann.Rev.Materl.Sci. 9 (1979) 151-189.
129   B.P.Kashyap and A.K.Mukherjee, Superplasticity, Ed.
      B.Baudelet and M.Suery, publ. CNRS, Paris (1985) Ch.4.
130   T.G.Langdon, Superplastic Forming of Structural Alloys,
      Ed. N.E.Paton and C.H.Hamilton, publ. TMS-AIME,
      Warrendale (1982) 27-40.
131   R.C.Gifkins, Superplastic Forming of Structural Alloys,
      Ed. N.E.Paton and C.H.Hamilton, publ. TMS-AIME,
      Warrendale (1982) 3-26.
132   A.E.Geckinli, Metal Sci. 17 (1983) 12-18.
133   W.Beere, Metal Sci. 10 (1976) 133-139.
134   J.P.Hirth and W.D.Nix, Acta Metall. 33 (1985) 359-368
135   J.H.Schneibel and P.M.Hazzledine, J.Materl.Sci. 18
      (1983) 562-570.
136   A.Arieli and A.K.Mukherjee, Materl.Sci and Eng. 45
      (1980) 61-70.
137   L.C.Samuelson, K.N.Melton and J.W.Edington,
      Acta Metall. 24 (1976) 1017-1026.
138   K.N.Melton and J.W.Edington, Metal Sci.J.
      7 (1973) 172-175.
139   H.Naziri, R.Pearce, M.H.Brown and K.F.Hale,
      J.Microsc.97 (1973) 229-238.

140   R.H.Bricknell and J.W.Edington, Acta Metall.
       25 (1977) 447-458.
141   J.W.Edington, K.N.Melton and C.P.Cutler,
       Prog.Materl.Sci 21 (1976) 67-170.
142   J.E.Bird, A.K.Mukherjee and J.E.Dorn,
       Quantitative Relation between Properties and
       Microstructure,, Ed. D.G.Brandon and A.Rosen,
       publ. Israeli Univ. Press, Jeruselam (1969) 255-341.
143   R.C.Gifkins, Metall.Trans. 7A (1976) 1225-1232.
144   L.K.L.Falk, P.R.Howell, G.L.Dunlop and T.G.Langdon,
       Acta Metall 34 (1986) 1203-1214.
145   R.B.Vastava and T.G.Langdon, Acta Metall. 27 (1979)
       251-257.
146   A.E.Geckinli and C.R.Barrett, J.Materl.Sci 11
       (1976) 510-521.
147   M.F.Ashby and R.A.Verrall, Acta Metall. 21
       (1973) 149-163.
148   J.R.Spingarn and W.D.Nix, Acta Metall 26 (1978)
       1389-1398.
149   W.Beere, Philos.Trans.Roy.Soc.Lon. A288 (1978)
       177-196.
150   A.Arieli, T.Kianuma and A.K.Mukherjee, Acta Metall.
       30 (1982) 1679-1688.
151   W.D.Nix, Superplastic Forming, Ed. S.P.Agrawal,
       publ. ASM, Metals Park (1985) 3-12.
152   M.Suery and B.Baudelet, Res Mechanica 2 (1981)
       163-170.
153   J.H.Gittus, Trans. ASME (1977) 244-251.
154   I.W.Chen, Acta Metall. 30 (1982) 1655-1664.
155   J.D.Eshelby, Proc.Roy.Soc.Lond. 241A (1957) 376-396.
156   I.W.Chen, Superplasticity, Ed. B.Baudelet and
       M.Suery, publ. CNRS Paris (1985) Ch.5.
157   E.W.Hart, Acta Metall. 15 (1967) 351-355.
158   J.D.Campbell, J.Mech.Phys.Solids 15 (1967) 359-370.
159   C.Rossard, Revue Metall. 63 (1966) 225-235.
160   D.A.Woodford, Trans. ASM 62 (1969) 291-293.
161   A.K.Ghosh and A.Ayres, Metall.Trans. 7A (1976)
       1589-1591.
162   D.H.Avery and J.M.Stuart, Surfaces and Interfaces II,
       Ed. J.J.Burke, N.L.Reid and V.Weiss, publ. Syracuse
       Univ. Press, New York (1968) 371.
163   D.L.Holt, Ultrafine Grain Metals, Ed. J.J.Burke
       and V.Weiss, publ. Syracuse Univ. Press, New York
       (1970) 355.
164   W.B.Morrison, Trans. TMS-AIME 242 (1968) 2221-2227.
165   M.J.Stowell, Metal Sci. 17 (1983) 92-98.
166   J.Lian and M.Suery, Mat.Sci.Technol. 2 (1986)
       1093-1098.
167   J.Lian and B.Baudelet, Mat.Sci. and Eng. 84 (1986)
       157-162.
168   M.J.Stowell, Metal Sci. 14 (1980) 267-272.
169   A.K.Mukherjee, Grain Boundaries in Engineering
       Materials, publ. Claitor Publishing, Baton Rouge
       (1975) 93-105.
170   A.K.Ghosh and C.H.Hamilton, ICSMA 5, Ed. P.Hassen,
       V.Gerold and G.Kostorz, publ. Pergamon Press,
       London (1979) 905-911.

171   J.Pilling and N.Ridley, 3rd. Int. Aluminium-Lithium
Conf. Ed. C.Baker, P.J.Gregson, S.J.Harris and
C.J.Peel, publ. Institute of Metals, London
(1985) 184-190.

172   D.A.Miller and T.G.Langdon, Trans JIM. 21 (1980)
123-125.

173   A.H.Chokshi, J.Materl.Sci.Lett.5 (1986) 144-146.

174   C.C.Bampton and J.W.Edington, Metall.Trans. 13A
(1982) 1721-1727.

175   T.Chandra, J.J.Jonas and D.M.R.Taplin, J.Materl.Sci.
13 (1978) 2380-2384.

176   G.L.Dunlop, E.Shapiro, D.M.R.Taplin and J.Crane,
Metall.Trans. 4A (1973) 2039-2044.

177   S.Sagat, P.Blenkinsop and D.M.R.Taplin,
J.Inst.Metals 100 (1972) 268-274.

178   J.Belzunce and M.Suery, Scripta Metall. 15 (1981)
895-898.

179   C.W.Humphries and N.Ridley, J.Materl.Sci. 13 (1978)
2477-2482.

180   S.A.Shei and T.G.Langdon, J.Materl.Sci. 13 (1978)
1084-1092.

181   C.H.Caceres and D.S.Wilkinson, Acta Metall. 32
(1984) 415-434.

182   W.B.Morrison, Trans, ASM 61 (1968) 423-434.

183   D.W.Livesey and N.Ridley, J.Materl.Sci. 13 (1978)
825-832.

184   M.M.Ennis, Superplastic Forming of Structural
Alloys, Ed. N.E.Paton and C.H.Hamilton,
publ. TMS-AIME, Warrendale (1982) 392.

185   Y.Ito and A.Hasegawa, Titanium '80, publ. TMS-AIME,
Warrendale (1980) 983-992.

186   G.Gurewitz, N.Ridley and A.K.Mukherjee, ICF Symposium
and Fracture, Paper 5a-6, Beijing, China (1983).

187   M.T.Cope, Ph.D. Thesis, University of Manchester (1984).

188   H.Ishikawa, D.G.Bhat, F.A.Mohamed and T.G.Langdon,
Metall.Trans 8A (1977) 523-525.

189   M.M.I.Ahmed, F.A.Mohamed and T.G.Langdon,
J.Materl.Sci. 14 (1979) 2913-2917.

190   D.A.Miller and T.G.Langdon, Metall.Trans 9A (1978)
1688-1690.

191   D.W.Livesey and N.Ridley, J.Materl.Sci. 17 (1982)
2257-2266.

192   N.Ridley and J.Pilling, Superplasticity, Ed.
B.Baudelet and M.Suery, publ. CNRS Paris (1985) Ch.8.

193   M.Suery, Superplasticity, Ed. B.Baudelet and M.Suery,
publ. CNRS Paris (1985) Ch.9.

194   B.P.Kashyap and A.K.Mukerjee, Res Mechanica 17
(1986) 293-355.

195   M.J.Stowell, Superplastic Forming of Structural
Alloys, Ed. N.E.Paton and C.H.Hamilton, publ.
TMS-AIME, Warrendale (1982) 321-336.

196   J.Pilling and N.Ridley, Res Mechanica 23 (1988)
31-63.

197   D.W.Livesey and N.Ridley, Met.Sci.16 (1982) 563-568.

198   C.C.Bampton and R.Raj, Acta Metall. 30 (1982)
2043-2053.

199   P.J.Mescheter, R.J.Lederich and P.S.Rao,
Sripta Metall. 18 (1984) 833-836.

200 A.C.F.Cocks and M.F.Ashby, Metal Sci. 14 (1980)
    395-402.
201 A.C.F.Cocks and M.F.Ashby, Metal Sci. 16 (1982)
    465-474.
202 A.C.F.Cocks and M.F.Ashby, Progress in Materl.Sci.
    27 (1982) 189-244.
203 D.Hull and D.Rimmer, Philos.Mag. 4 (1959) 673-687
204 M.V.Speight and J.E.Harris, Metal Sci. 1 (1967)
    83-85
205 M.V.Speight and W.Beere, Metal Sci. 9 (1975)
    190-191.
206 W.Beere and M.V.Speight, Metal Sci. 12 (1978)
    172-176.
207 I.W.Chen and A.S.Argon, Acta Metall.29 (1981)
    1759-1768.
208 R.Raj and M.F.Ashby, Acta Metall.23 (1975) 653-666.
209 G.H.Edward and M.F.Ashby, Acta Metall. 27 (1979)
    1505-1518.
210 D.A.Miller and T.G.Langdon, Metall.Trans.10A (1979)
    1869-1873.
211 J.Pilling, D.W.Livesey, J.B.Hawkyard and N.Ridley,
    Metal Sci. 18 (1984) 117-122.
212 A.K.Chokshi and T.G.Langdon, Acta Metall. 35 (1987)
    1089-1101
213 A.Needleman and J.R.Rice, Acta Metall. 28 (1980)
    1315-1332.
214 J.W.Hancock, Metal Sci. 10 (1976) 319-325.
215 J.R.Rice and D.M.Tracey, J.Mech.Phys.Solids 17
    (1969) 201-217.
216 B.Budiansky, J.W.Hutchinson and S.Slutsky,
    Mechanics of Solids, Ed. H.G.Hopkins and
    M.J.Sewell, publ. Pergamon Press, Oxford (1982) 13.
217 M.J.Stowell, D.W.Livesey and N.Ridley, Acta Metall.
    32 (1984) 35-42.
218 J.Pilling and N.Ridley, Acta Metall. 34 (1986)
    669-679.
219 N.Ridley and D.W.Livesey, Res Mechanica Lett. 1
    (1981) 73-78.
220 N.Ridley, D.W.Livesey and A.K.Mukherjee,
    J.Materl.Sci. 19 (1984) 1321-1332.
221 W.J.D.Patterson and N.Ridley, J.Materl.Sci. 16
    (1981) 457-466.
222 L.B.Duffy, M.Sc. Thesis, University of Manchester (1982).
223 J.Pilling, Materl.Sci. and Technol. 1 (1985) 461-466.
224 J.M.Storey, J.I.Petit, D.J.Lege and B.L.Hazard,
    Superplasticity in Aerospace, Ed.
    R.Pearce and L.Kelly, publ. Cranfield
    Institute of Technology, Bedford (1986) 67-104.
225 R.A.Tait and D.M.R.Taplin, Scripta Metall. 13 (1979)
    77-82.
226 A.Varloteaux and M.Suery, Superplasticity in
    Aerospace - Aluminium, Ed. R.Pearce and L.Kelly,
    publ. Cranfield Institute of Technology, Bedford
    (1986) 55-66.
227 H.Ahmed and R.Pearce, Superplasticity in Aerospace -
    Aluminium, Ed. R.Pearce and L.Kelly, publ. Cranfield
    Institute of Technology, Bedford (1986) 146-159.

228 C.C.Bampton, A.K.Ghosh and M.W.Mahoney, Superplasticity in Aerospace - Aluminium, Ed. R.Pearce and L.Kelly, publ. Cranfield Institute of Technology, Bedford (1986) 1-35.

229 H.E.Evans, Mechanisms of Creep Fracture, publ. Elsevier Applied Science, New York.

230 T.J.Chang, K.I.Kagawa, J.R.Rice and L.B.Sills, Acta Metall. 27 (1979) 265-284.

231 P.Shariat, R.B.Vastava and T.G.Langdon, Acta Metall. 30 (1982) 285-296.

232 M.H.Yoo and H.Trinkhaus, Metall.Trans. 14A (1983) 547-561.

233 N.Ridley, D.W.Livesey and A.K.Mukherjee, Metall.Trans. 15A (1984) 1443-1450.

234 R.Raj, Acta Metall. 26 (1978) 995-1006.

235 J.S.Wang, J.J.Stephans and W.D.Nix, Acta Metall. 33 (1985) 1009-1021.

236 M.Suery and B.Baudelet, Superplastic Forming of Structural Alloys, Ed. N.E.Paton and C.H.Hamilton, publ. TMS-AIME, Warrendale (1982) 105-127.

237 H.M.Shang and M.Suery, Metal Sci. 18 (1984) 143-152

238 J.Pilling, B.Geary and N.Ridley, ICSMA 7, Ed. H.J.McQueen and J.P.Bailon, publ. Pergamon Press, Oxford, (1985) 823-828.

239 A.K.Ghosh, Deformation of Polycrystals - Mechanics and Microstructure, Ed. N.Hansen, Publ. Riso National Laboratories, Denmark (1981) 277-282.

240 N.Ridley, C.W.Humphries and D.W.Livesey, ICSMA 4, publ. ENSMIM, Nancy (1976) 433-437.

241 R.Sawle, Superplastic Forming of Structural Alloys, Ed. N.E.Paton and C.H.Hamilton, publ. TMS-AIME, Warrendale (1982) 307-317.

242 A.J.Barnes, Superform Metals, Worcester (1983).

243 Superform Metals Ltd. publication #M307, Worcester UK.

244 W.C.Zhang, R.D.Wood and O.C.Zienkiewicz, Al-Technology, Ed. T.Sheppard, publ. Institute of Metals, London (1986) 635-640.

245 S.Y.Quan and Z.Juan, Materl.Sci and Eng. 84 (1986) 111-125.

246 A.R.Ragab, Met. Technol.10 (1983) 340-348.

247 G.C.Cornfield and R.H.Johnson, Int.J.Mech.Sci.12 (1970) 479-490.

248 D.L.Holt Int.J.Mech.Sci.12 (1970) 491-497.

249 W.Johnson, T.Y.M.Al-Naib and J.L.Duncan, J.Inst. Metals 100 (1972) 45-50.

250 F.Jovane, Int.J.Mech.Sci. 10 (1968) 403-427.

251 M.F.Ashby, Metall.Trans. 14A (1985) 1755-1769.

252 J.R.Williamson, Superplastic Forming of Structural Alloys, Ed. N.E.Paton and C.H.Hamilton, publ. TMS-AIME, Warrendale (1982) 291-306.

253 E.D.Weisert and G.W.Stacher, Superplastic Forming of Structural Alloys, Ed. N.E.Paton and C.H.Hamilton, publ. TMS-AIME, Warrendale (1982) 273-289.

254 U.S Patent 3927817, Rockwell International ltd. (1975).

255 C.H.Hamilton, Titanium Science and Technology, Ed. R.I.Jaffe and R.M.Burte, publ. Plenum Press (1973) 625-647.

256   G.Garmong, N.E.Paton and A.S.Argon, Metall.Trans. 6A (1975) 1269-1279.

257   B.Derby and E.R.Wallach, Metal Sci. 16 (1982) 49-56.

258   B.Derby and E.R.Wallach, Metal Sci. 18 (1984) 427-431.

259   D.S.Wilkinson and M.F.Ashby, Acta Metall. 23 (1975) 1277-1285.

260   J.Pilling, Materl.Sci. and Eng. 100 (1988) 137-144.

261   S.Reusswig, R.Gleichmann, P.G.Zielinski, D.G.Ast and R.Raj, Acta Metall. 32 (1984) 1553-1560.

262   N.Furishiro and S.Hori, Titanium '80, Ed. H.Kimura and O.Izumi, publ. TMS-AIME (1980) 1067-1071.

263   C.I.Smith, B.Norgate and N.Ridley, Scripta Met, 8 (1974) 159-164.

264   G.Tober and S.Elze, Advanced Joining of Metallic Materials, AGARD CP398 (1986) Ch.11.

265   R.G.Wing and M.Newnhan, in Diffusion Bonding, Ed. R.Pearce Cranfield School of Industrial Science, (1987) 159-172.

266   D.Stephen and S.J.Swadling, Advanced Joining of Metallic Materials, AGARD CP398 (1986) Ch.7.

267   D.Stephen, Designing with Titanium, Institute of Metals, London (1986) 108-115.

268   T.S.Baker and P.G.Partridge, in Titanium Science and Technology, Ed. G.Lutjering, U.Zwicker and W.Bunk, DGM, Munich, 2 (1985) 861-868.

269   V.Alerja, in Titanium Science and Technology, Ed. G.Lutjering, U.Zwicker and W.Bunk) DGM, Munich, 2 (1985) 933-938.

270   Z.A.Munir, Weld.J.Res.Supp.62 (1983) 333s-336s.

271   T.S.Baker and P.G.Partridge, in Diffusion Bonding, Ed. R.Pearce, Cranfield School of Industrial Science, (1987) 73-89.

272   J.Harvey, P.G.Partridge and A.M.Lurshay, Materl.Sci. and Eng. 79 (1986) 191-199.

273   J.Harvey, P.G.Partridge and C.L.Snooke, J.Materl.Sci. 20 (1985) 1009-1014.

274   R.V.Sharples and I.A.Bucklow, Res.Rep #307/1986, The Welding Institute, Cambridge UK. (1986).

275   A.E.Dray and E.R.Wallach, in Diffusion Bonding, Ed. R.Pearce, publ. Cranfield School of Industrial Science, (1987) 125-128.

276   N.Ridley, J.Pilling, A.Tekin and Z.X.Guo, Diffusion Bonding, Ed. R.Pearce, publ. Cranfield School of Industrial Science, (1987) 129-142.

277   T.D.Byun and R.B.Vastava, Welding, Bonding and Fastening, Ed.J.D.Buckley and B.A.Stein, NASA Langley Research Centre, Hampton Va. (1984) 231-245.

278   T.D.Byun and P.Yavari, Superplasticity in Aerospace - Aluminium, Ed. R.Pearce and L.Kelly, Cranfield, (1985) 285-294.

279   J.Pilling and N.Ridley, Materl.Sci. and Technol. 3 (1987) 353-359.

280   S.H.Reichman and J.W.Smythe, Int J. Powder Met. 6 (1970) 65-74.

281   R.G.Menzies, J.W.Edington and G.J.Davies, Metal Sci. 15 (1981) 210216.

282   Y.Bienvenu, T.Massart, L.Van Wouw, M.Jeandin and
      A.Morrison, <u>Diffusion Bonding</u>, Ed. R.Pearce,
      Cranfield School of Industrial Science (1987) 33-43.
283   Y.Hiroki and B.J.Pletka, private communication.
284   T.Massart, Y.Bienvenu and A.Morrison,
      <u>Diffusion Bonding</u>, Ed. R.Pearce, Cranfield
      School of Industrial Science (1987) 91-101.
285   W.T.Chandler, A.K.Ghosh and M.Mahoney, J. Spacecr.
      Rockets, 21 (1984) 61-67.

BIBLIOGRAPHY

[A] Superplasticity   —   K.A.Padmanabhan and G.J.Davies, Springer-Verlag, Berlin, 1980.

[B] Superplastic Forming of Structural Alloys — Ed. N.E.Paton and C.H.Hamilton TMS-AIME, Warrendale, Pa., 1982.

[C] Superplasticity in Aerospace - Aluminium — Ed. R.Pearce and L.Kelly, Cranfield Institute of Technology, Cranfield, 1985.

[D] Superplasticite — Ed. B.Baudelet and M.Suery, Centre National de Recherche Scientifique, Paris, 1985.

[E] Superplastic Forming — Ed. S.P.Agrawal, ASM, Metals Park, OH., 1986.

[F] Superplasticity — AGARD-LS-154, NATO/OTAN, Nuelliy-sur-Seine, 1987.

[G] Diffusion Bonding — Ed. R.Pearce, Cranfield School of Industrial Science, Cranfield, 1987.

[H] Superplasticity in Aerospace — Ed. H.C.Heikkenen and T.R.McNelley TMS-AIME, Warrendale, Pa. 1988.

[I] Superplasticity and Superplastic Forming — Ed. C.H.Hamilton and N.E.Paton TMS-AIME, Warrendale, Pa. 1989.

INDEX